Reproduction and Biopolitics

The central theme of this volume is the notion of 'irrational reproduction': the ways in which women's and couples' reproductive choices and practices are deemed 'irrational' or 'irresponsible' because they result in the 'wrong number' of children. In a global context of declining fertility, population policies have shifted to a neoliberal register, which, despite local differences, includes both the deepening of economic and social inequalities and the intensification of rights discourses applied to the unborn. Inspired by Foucault's theories on biopolitics and biopower and by a long tradition of feminist anthropological studies on reproduction, the ethnographically based papers collected in this volume address the following crucial questions: How does the notion of 'irrational' reproduction emerge and play out in diverse socio-political contexts and what forms of subjectivities and resistance does it generate? How does the 'threat' of too few or too many children, itself constructed through expert knowledge of statistics and political concerns over the size of different ethnic populations or classes, justify and support different biopolitical projects? And how does the increasing privatization of healthcare and the dismantling of welfare states affect reproductive practices and decisions on the ground in the global North and South?

This book was originally published as a special issue of *Anthropology & Medicine.*

Silvia De Zordo is Postdoctoral Beatriu de Pinós Fellow at the University of Barcelona, Spain. She has over ten years of research experience on health, gender and reproductive politics in Latin America and Europe. Her current research focuses on abortion and conscientious objection in Europe (Italy, Spain and UK).

Milena Marchesi is Lecturer at the University of Massachusetts Amherst, USA. Her research focuses on reproductive, family and immigration politics in Italy.

Reproduction and Biopolitics

Ethnographies of Governance, 'Irrationality' and Resistance

Edited by
Silvia De Zordo and Milena Marchesi

Routledge
Taylor & Francis Group
LONDON AND NEW YORK

First published 2015
by Routledge
2 Park Square, Milton Park, Abingdon, Oxon, OX14 4RN, UK

and by Routledge
711 Third Avenue, New York, NY 10017, USA

Routledge is an imprint of the Taylor & Francis Group, an informa business

British Library Cataloguing in Publication Data
A catalogue record for this book is available from the British Library

ISBN 13: 978-1-138-80322-0

Typeset in Times New Roman
by RefineCatch Limited, Bungay, Suffolk

Publisher's Note
The publisher accepts responsibility for any inconsistencies that may have arisen during the conversion of this book from journal articles to book chapters, namely the possible inclusion of journal terminology.

Contents

Citation Information vii

Notes on Contributors ix

1. Introduction. Ethnography and biopolitics: tracing 'rationalities' of reproduction across the north–south divide
Elizabeth L. Krause and Silvia De Zordo 1

2. Irrational non-reproduction? The 'dying nation' and the postsocialist logics of declining motherhood in Poland
Joanna Mishtal 17

3. Reproducing Italians: contested biopolitics in the age of 'replacement anxiety'
Milena Marchesi 35

4. Islamic logics, reproductive rationalities: family planning in northern Pakistan
Emma Varley 53

5. Programming the body, planning reproduction, governing life: the '(ir-) rationality' of family planning and the embodiment of social inequalities in Salvador da Bahia (Brazil)
Silvia De Zordo 71

6. The right to have a family: 'legal trafficking of children', adoption and birth control in Brazil
Andrea Cardarello 89

7. Reproductive governance in Latin America
Lynn M. Morgan and Elizabeth F.S. Roberts 105

Index 119

Citation Information

The chapters in this book were originally published in the *Anthropology & Medicine*, volume 19, issue 2 (August 2012). When citing this material, please use the original page numbering for each article, as follows:

Chapter 1

Introduction. Ethnography and biopolitics: tracing 'rationalities' of reproduction across the north–south divide
Elizabeth L. Krause and Silvia De Zordo
Anthropology & Medicine, volume 19, issue 2 (August 2012) pp. 137–151

Chapter 2

Irrational non-reproduction? The 'dying nation' and the postsocialist logics of declining motherhood in Poland
Joanna Mishtal
Anthropology & Medicine, volume 19, issue 2 (August 2012) pp. 153–169

Chapter 3

Reproducing Italians: contested biopolitics in the age of 'replacement anxiety'
Milena Marchesi
Anthropology & Medicine, volume 19, issue 2 (August 2012) pp. 171–188

Chapter 4

Islamic logics, reproductive rationalities: family planning in northern Pakistan
Emma Varley
Anthropology & Medicine, volume 19, issue 2 (August 2012) pp. 189–206

Chapter 5

Programming the body, planning reproduction, governing life: the '(ir-) rationality' of family planning and the embodiment of social inequalities in Salvador da Bahia (Brazil)
Silvia De Zordo
Anthropology & Medicine, volume 19, issue 2 (August 2012) pp. 207–223

Chapter 6

The right to have a family: 'legal trafficking of children', adoption and birth control in Brazil
Andrea Cardarello
Anthropology & Medicine, volume 19, issue 2 (August 2012) pp. 225–240

Chapter 7

Reproductive governance in Latin America
Lynn M. Morgan and Elizabeth F.S. Roberts
Anthropology & Medicine, volume 19, issue 2 (August 2012) pp. 241–254

Please direct any queries you may have about the citations to
clsuk.permissions@cengage.com

Notes on Contributors

Andrea Cardarello holds a PhD in Anthropology from l'Université de Montréal, and has taught anthropology and sociology at universities in Canada and Brazil. She has done research on issues related to kinship, adoption, foster care, poverty, children's rights and immigration.

Silvia De Zordo is Postdoctoral Beatriu de Pinós Fellow at the University of Barcelona, Spain. She has over ten years of research experience on health, gender and reproductive politics in Latin America and Europe. Her current research focuses on abortion and conscientious objection in Europe (Italy, Spain and UK).

Elizabeth L. Krause is Professor of Anthropology at the University of Massachusetts Amherst, USA and author of *Unraveled: A Weaver's Tale of Life Gone Modern* (2009) and *A Crisis of Births: Population Politics and Family-Making in Italy* (2005).

Milena Marchesi is Lecturer at the University of Massachusetts Amherst, USA. Her research focuses on reproductive, family and immigration politics in Italy.

Joanna Mishtal is Assistant Professor of Anthropology at the University of Central Florida, USA. Her current research, situated in Ireland and the European Union, examines the intersection of medicine, religion, and policy through the lens of the use of conscience based objection in reproductive healthcare.

Lynn M. Morgan is Mary E. Woolley Professor of Anthropology at Mount Holyoke College, Massachusetts, USA and author of *Icons of life: A Cultural History of Human Embryos* (2009).

Elizabeth F.S. Roberts is Associate Professor of Anthropology at the University of Michigan, USA and author of *God's Laboratory: Assisted Reproduction in the Andes* (2012). She is currently working on a project about environmental health science and epigenetics in Mexico City.

Emma Varley is Assistant Professor of Anthropology at Brandon University, Manitoba, Canada. Her research addresses the impacts of conflict, structural inequality, and health sector reform for Safe Motherhood and Maternal, Newborn and Child Health (MNCH) services, programming, and policies in northern Pakistan.

Introduction. Ethnography and biopolitics: tracing 'rationalities' of reproduction across the north–south divide

Elizabeth L. Krause[a] and Silvia De Zordo[b]

[a]*Department of Anthropology, University of Massachusetts Amherst, Amherst, USA;*
[b]*Department of Anthropology, Goldsmiths – University of London, New Cross, London, UK*

This collection of papers assembles the work of international ethnographers plying their painstaking methods in different nooks of the world. Their projects collectively confront tyrannies with truths. Such endeavors are not easily accomplished, for the truth about reproductive politics that these papers expose has largely lingered in the shadows of family planning and its rationalist paradigm.

This special issue reveals how rationalities concerned with reproductive and sexual bodies arise and circulate over historic time and across social spaces. Despite wildly different contexts, remarkable similarities crosscut the 'global North' and 'global South'. The papers draw attention to the ways in which policies and practices differ in their specifics yet also unpredictably mirror one another in general ways. As these biopolitical rationalities change over time, new tactics, truths, and moral regimes emerge. In parallel, they transform subjectivities and foment strategies of negotiation, contestation, and resistance.

To offer intellectual glue, this introduction has three objectives: (1) to situate these papers in the context of biopolitics and clarify what has become a somewhat catch-all, hazy concept; (2) to specify the relevance of ethnography for enacting a 'genealogical method' in order to question assumptions about rational reproduction; and (3) to illuminate themes that emerge in an epistemic moment in which neoliberal ideologies of privatization and other forces of globalization influence social spaces in the 'global North' and 'South'.

One of the most salient global shifts in the past decades stems from trends of declining fertility. Population policies targeted at reducing births have waned. Given that those policies were promoted in the guise of modernity, development, and linear notions of progress (Greenhalgh 1995), one might imagine that debates over reproductive practices would have largely vanished. Instead, the tired neo-Malthusian key has shifted to a neoliberal register in which reproductive politics aim at defending, granting, and enhancing individual or collective rights and wellbeing. These include an intensification of the rights of the unborn and a deepening of economic and social inequalities. The papers collected in this issue attest to the myriad ways in which different biopolitical rationalities strive to discipline knowledge, embodied practices, and 'life as such,' meaning life understood as how it is 'lived through a body (not only through cells) and as a society (not only as species)' (Fassin 2009, 48). They also show how individuals struggle to satisfy their reproductive desires – to have, to accept, or to avoid having children – and to embrace, question, or contest expert interventions and surveillance. Paradoxes abound.

Biopolitics refreshed

The concept of biopolitics arguably finds its intellectual roots in Thomas Malthus's (1798) *An essay on the principle of population*, with a critical response some six decades later in Karl Marx's (1906[1867]) writings in *Capital* elaborating on the population–poverty relationship, and contemporary formulations a century later in the work of Michel Foucault (1990[1978]), particularly his *History of sexuality, Vol. 1*. In turn, contemporary scholars have been interested in the linkages between the construction of scientific problems and the ways in which governmental or biomedical practices elaborate on those defined problems (e.g., Horn 1994, 7; see also Fassin 2010, Greenhalgh 2003, Martin 2001[1987], Rapp 2000, Rose 2007). The objects of scholarship have undergone dramatic shifts toward examining what Ian Hacking described as 'the public life of concepts and the ways in which they gain authority' (Hacking 1990, 7 cited in Horn 1994, 7).

To lend perspective to the ways in which the science of family planning gained authority, and its old relationship to class conflict and inequality, it is worth turning briefly to some old and dusty intellectual history. Malthus, of course, used neither the term biopolitics nor biopower but his work was a precursor to both. He viewed high fertility as a sign of ignorance and moral bankruptcy and a justification for policies designed to give the poor their just desserts: starvation. He argued against the Poor Bill's stipulation to allow 'a shilling a week to every laborer for each child he has above three' (Malthus 1798, 42), reasoning that it would not result in the intended effect to improve the life of the poor but rather its opposite. He argued not for abstinence but for regulation, especially in the form of 'preventive checks', such as delayed marriage and celibacy before marriage (see Schneider and Schneider 1996, 18–22). Ultimately, people had to exercise moral restraint in order to counter the 'laws of nature', a term used to describe 'the passion between the sexes', which for Malthus was immutable over time (Malthus 1798, 40). Clearly, attempts at defining moral regimes through governance have a long history.

It is worth pointing out that both Malthus and Marx were population pessimists in that both viewed big populations in a negative light against Adam Smith's wealth of nations or, say, Mussolini's urge to stimulate population growth. Both viewed high fertility and widespread pauperism as depressing wages. They saw inverse relationships between wages and population growth. The similarities ended there, to put it simply. When it came to the issue of surplus population – or too many people – Malthus attributed the causes to the eternal law of population whereas Marx emphasized the historical laws of capital. Regarding public assistance, Malthus viewed the poor laws as hampering happiness among the commoners

(Malthus 1798, 29) whereas Marx described the assault on these laws as a war on the proletariat. Indeed, Marx (1906, 703–11) viewed surplus population as necessary for capitalism, keenly criticizing efforts to retain and exploit reserve armies. Class interests in managing populations were thus revealed.

The concept of '*biopolitics of the population*' (Foucault 1990[1978], 139) elaborates on both of these major works albeit in a very different historical moment and style of intellectual production. Foucault demarcated a shift in the exercise of power in Western societies. The old form privileged the sovereign ruler in his role to determine death; the new form witnessed a diffuse form aimed at administering life. In Foucault's theories of State political activity, the quality and quantity of those inhabitants living in a given territory became essential. Attention came to focus on the ways in which modern forms of government counted and managed populations through a variety of techniques to 'know' the aggregate. Turning the population into an 'object of knowledge' ultimately then allowed for justifications to enact 'normalizing interventions', as Susan Gal and Gail Kligman (2000, 19) have observed. Modern power has certain characteristics: it 'exerts a positive influence on life that endeavors to administer, optimize, and multiply it, subjecting it to precise controls and comprehensive regulations' (Foucault 1990[1978], 137, also cited in Rabinow 1984, 259, discussed in Gal and Kligman 2000, 17–20 and Oksala 2007, 82–3).

Thus, biopolitics are never stagnant. They are the State in action. Here, the State is not merely a unitary, static thing but a set of practices involving all of its legislators, all of its inhabitants. Tactics of biopolitics bounce like a video game character off two crucial poles – 'disciplines of the body' and 'regulations of the population' – for it is here that a certain 'power over life' has been and continues to be organized and deployed (Foucault 1990[1978], 139). These particular techniques cannot be taken-for-granted; their limits can never be predicted. On the one hand, biopolitics allows for conceiving power as not merely top down but as diffuse, such as when individuals become subject to norms of behavior and may internalize those norms yet also modify them as they do so. On the other hand, biopolitics reminds us of the importance of expert practices such as statistics in shaping policies and exercising power geared at the 'large-scale phenomena of population' (Foucault in Rabinow 1984, 260).

Foucault's work has had a powerful influence on feminist anthropologists and others who have sought to connect large-scale global influences with local practices and experiences, often with keen attention to manifestations of 'stratified reproduction', as Faye Ginsburg and Rayna Rapp (1995) note in their volume *Conceiving the new world order*. Indeed, the debates, practices, and policies surrounding acceptable terms of reproduction illuminate a politics of embodiment writ large (after Lock and Farquhar 2007). In modern nation-statecraft, as Gal and Kligman (2000, 15) have suggested for Eastern Europe, reproduction is not simply limited to the private domain but plays a central role in the spheres of politics, the State, and civil society.

The papers collected in this special issue enrich this literature with a comparative perspective. From post-socialist Europe to Western Europe and the hinterlands of Asia, to the cores and peripheries of Latin America, distant places reveal patterns of paradox. Particularly stunning is that in these diverse contexts, women, as the main target of reproductive policies, are often stigmatized regardless of their behaviors: for postponing motherhood, for embracing childlessness, or for having children. Even so, women are not 'docile bodies'. On the ground, biopolitics rarely go unnoticed and uncontested.

Ethnographies of biopolitics

The painstaking and often heartbreaking observations contained in these papers reveal concrete examples of upsetting the tyranny of globalizing discourses (after Foucault 1980). Illuminations result from thorough, meticulous, and conscientious research. Such work consumes time and energy. It is not easily bounded, nor is it easily concluded. The centrality of reproduction continues to bear profoundly on the ongoing projects of nation-building and subject-making, and to remind of technologies concerned above all with the management of life. Countering such modern forms of power entails, as Foucault predicted, 'a painstaking rediscovery of struggles together with the rude memory of their conflicts' (Foucault 1980, 83). The lived experiences revealed in these ethnographic accounts, from points North and South, remind us that women around the globe continue to experience conflict related to their reproductive practices, and that more than babies are at stake. Citizenship, personhood, rights – the most basic elements of being human coalesce in the realm of fertility politics. This is biopolitics in all its glory. The comparative context exposes all matter of contradictions, and yet the impulse to manage life persists.

One imagines Silvia De Zordo sitting in clinic hallways, listening attentively to working-class black women in Brazil, their memories heavy with neo-Malthusian 'gifts' such as tied tubes in exchange for the 'right' to vote. One sees Joanna Mishtal carrying out interviews with 55 women reeling from Poland's neoliberal restructuring, sharing raw recollections of job discrimination and fears of pregnancy. In Italy, one follows Milena Marchesi through the streets of Milan marching in a mock funeral for the unfertilized egg in protest of a new fetal burial law and listening to feminists grapple with attacks on their personhood against a backdrop of cries of a 'country dying from low birth rates'. One senses Andréa Cardarello's entanglements in Brazil with officials who morally and judicially sanction seizing children from their downtrodden mothers and fathers in the name of legal 'adoptions' geared at fighting poverty. In Northern Pakistan, one pictures Emma Varley in a battlefield of new attempts to deliver Islamic perspective on family planning.

The evidence resonates. The end of history has hardly arrived when it comes to reproduction (cf. Fukuyama 1992). Some major trends in population might have tempted one to think that population is a finished topic. Fertility is declining in most parts of the world. Overpopulation doomsayers have revised their alarms. The United Nations has lowered its prediction of peak world population to 9.2 billion in 2050. Predictions are less dire than back in the day of Paul Ehrlich's (1968) *Population bomb*. Actual reproductive practices will bear on whether global population reaches or exceeds that prediction. If that's all fertility were about, the story would have been over long ago. The tropes of population gloom persist. The biopolitics of reproduction rage. They inspire resistance.

Moments of contestation are especially vivid because these ethnographers merge large-scale with small-scale phenomena. They track back and forth between large-scale phenomena – statistics, discourses, policies, and 'expert' interventions – and small-scale consequences on those who are the subjects of biopolitics. This captures the essence of genealogical research. These papers lend themselves to the power of a social-genealogical method in two senses of the term, one literal, the other figurative. First, the literal: they are genealogical in that their point of departure results from changes in patterns related to family descent. Where branches on the genealogical trees are shrinking, trouble brews. Where branches on the genealogical tree are growing, trouble brews. Conclusion: trouble brews. Reproducers are subject to State surveillance and criticism whether in the lowest-low fertility contexts of Europe, relatively new low-fertility contexts of Latin America, or high-fertility contexts of Islamic Asia. The ongoing gaze appears to be pre-programmed into the workings of biopower.

Second, the figurative: these ethnographers merge subjugated with erudite forms of knowledge (after Foucault 1990[1978]). The targeted subjects share a backdrop of demographic science, considered a highly legitimate and authorized form of expert knowledge, particularly when compared with quirky, unpredictable, narratives collected in marginal spaces and concerning messy, even embarrassing, topics atypical to science.

Furthermore, these papers contain compelling evidence for why ethnographers must strive to avoid the empiricist fallacy. As Paul Willis and Mats Trondman (2000:12) remind in their manifesto to ethnography, 'the "nitty gritty" of everyday life cannot be presented as raw, unmediated data' (Willis and Trondman 2000, 12). Those working in the sphere of fertility are particularly susceptible to falling victim to this fallacy. Researchers are bombarded with numbers that appear hard and real, numbers that claim to display the truth – whole and unmediated – through startling statistics. Reproductive practices that manifest within specific numeration contexts lend themselves to facile cause-and-effect explanations. Caution should be the guide.

Poorer societies historically had high fertility rates, yet all kinds of economic/modernization conditions existed at the time of demographic transition. The Princeton European Fertility Project's historical demographers disproved their hypothesis that modernization drove the onset of fertility decline (Coale and Watkins 1986). In the end, they pointed to cultural factors. Painstaking ethnographic work pushes cultural complexity. It is not the empirical facts themselves that cause fertility to decline or rise. It is people's particular responses and local expectations that matter, and how those intersect with large-scale exercises of power. Taken together, these papers allow us to wipe clear the wooly concept of biopolitics and remind us of its relevance as a tool to analyze forms of power that write themselves onto and into bodies.

Common and uncommon themes

The papers in this issue give much-needed attention to the 'materiality of population' as Jennifer Johnson-Hanks calls for in an *Annual Review of Anthropology* piece on 'Demographic transitions and modernity'. There, Johnson-Hanks (2008, 8) discusses 'vibrant literatures at the margins of population processes, both on the discourses of population and biopower/biopolitics', citing authors as diverse as Agamben (1998), Foucault (1990[1978], 2004), Greenhalgh and Winckler (2005), Krause (2005), Paxson (2004), and Rose (1996, 2007), and suggesting that what these literatures need to enhance them is direct 'engagement with the study of population size, structure, and rates of change, that is, with the materiality of population, and not only the discourses about it'.

These case studies ground the abstract concept of biopolitics. Taken individually, there are wide variations in the degree to which a biopolitical scaffolding frames the analysis. Taken as a whole, they provide examples of the ways in which discourses, practices, and policies about population manifest in people's lives in very gendered ways, how people and actors come to negotiate these forces, and why it all matters.

Exposures

What does a focus on biopolitics expose about how fertility-related discourses, practices, and policies manifest in people's lives? Discipline and surveillance emerge as overarching themes common to many of the case studies in this special issue. This may come as a surprise given the general shift in family planning from population-control to empowerment and rights. On a general scale, the disciplining measures and related surveillance of gendered and sexual bodies aim to get people to conform to norms related to contraception and reproduction across geopolitical contexts. At a specific level, the targets and strategies expose a surprising range of overlapping consequences: how women and couples experience stigma for not adhering to narrow norms; how accusations of irrational behavior take shape; and how new stakes for 'reasonable' and 'responsible' reproduction are cast.

Mishtal's spotlight on mechanisms of discipline and surveillance in Poland contrasts the socialist era with the postsocialist moment, exploring what she defines as 'the postsocialist logics of motherhood'. She compares biopolitics under socialism with the biopolitics of neoliberalism. Under State socialism, there existed generous State welfare provisions. In the neoliberal era, many of those provisions were eliminated in ways that hit women and would-be mothers particularly hard, reducing family and maternity benefits, and largely abolishing worker job security in favor of deregulated markets. Since then, experts have come to place center stage a demographic crisis of low fertility and aging. To manage the situation, postsocialist policies have embraced a Catholic morality and attacked reproductive rights: criminalizing abortion, limiting contraception, and eliminating sex education; meanwhile, discourses reverberate that accuse women of being irrational non-reproducers.

What emerges from this paper and others is the whirl of stigma related to the number of children women have, regardless of socioeconomic status or ethnic identity. Mishtal's paper points out an important shift in biopolitical rationalities operating in low-fertility European countries. Here, 'reproductive stigma' does not affect only the poor – immigrants from rural areas in the past, immigrants from the 'global South' in the present – or minority groups. Roma women are particularly stigmatized, in Poland, for their high fertility rates and perceived as a threat to the nation's social and religious cohesion and identity. Moreover, stigma extends even to middle-class Catholic women. These Polish women find themselves in a double bind of stigmatizing forces: If they do not have children or postpone motherhood, the State and the Catholic Church labels them as selfish, anti-patriotic, and anti-Christian; if they have more than one child, their co-workers view them as irrational, and their employers discriminate against them.

Similarly, in Italy, experts from demographers and physicians to conservative politicians label and thus stigmatize Italian women as 'irrational' if they postpone motherhood to prioritize their professional training or jobs, while migrant women are stigmatized as 'irrational' for having too many (foreign) children, but also for having too many abortions. The rigid politics of life operating in Italy supported by the Catholic Church and sympathetic politicians, defends the 'life' and the rights of the embryo and the ideal Catholic family at all costs. As a result, women who do not have children or who postpone motherhood are stigmatized, as are infertile women and couples who confront a restrictive law on medically assisted technologies, which excluded single women and same-sex couples.

Shifting to points South, the price of stigma literally has cost some poor Brazilians their children, given up for international adoption mostly to European (and 40% to Italian) couples. Cardarello cogently shows how human rights discourse can be selective, patronizing, and detrimental. During the 1990s and 2000s, such framings were turned against impoverished parents to defend the rights of poor children. Legal professionals, judges, and social workers put forth arguments about rights to have a proper family and to have access to health and education, and pitted the rights of children against those of parents and families. Dire poverty became a reason to designate parents as unfit. Morality has been used and abused, Cardarello argues, to legitimize and justify the irregular international adoption of these poor children, whose rights and wellbeing would not be granted, according to the judges and lawyers involved in this case of 'legal child trafficking', by their 'disorganized' families. Poor children's parents were labeled, in judges' and lawyers' discourses, as 'irresponsible' if they left their children to 'strangers' – often relatives or family friends – while they worked far from their homes, and 'immoral' if they had divorced or intended to do so. Their unstable economic conditions also contributed to making them appear in the eyes of the courts as 'bad' parents and to justify the revocation of their parental rights.

Cardarello argues that this case shows the 'universalizing posture of the law' (Bourdieu 1987), but also the violence that the State and its institutions can exercise on groups of citizens in the name of the defense of others' rights, including those of children. In fact, the parental rights of the hundreds of couples whose children were illegally adopted were often revoked without their consent. Not only were most of these adults illiterate or semi-illiterate, but they had 'faith in the legal system' (Cardarello, this issue), so they signed the documents the social workers or the lawyers asked them to sign without necessarily reading them. Those who dared to contest the judges' decision when their children were taken away were threatened or detained.

Furthermore, Cardarello's paper reveals the strength and persistence of neo-Malthusian ideas about family size to inspire and orient harsh reproductive and family policies. In the 1980s, the most common form of contraception in Brazil was sterilization. A 1991 Parliamentary Commission found that the practice of doing the procedure without formal consent was rampant, particularly among poor and indigenous women. Some employers even required women to present a certificate of sterilization before hiring them. Despite the knowledge of stratified sterilization, the rates increased from 31% in 1986 to 40% by 1996 with a marked decline in the use of the pill during this same period.

This increase, as Cardarello and De Zordo point out, was primarily the long-term result of birth control campaigns carried out during the military dictatorship (1964–84) by private family planning institutions not only in Brazil, but also in many other Latin American countries, funded in large part by international and mainly US-based organizations. In this period, Brazilian physicians provided tubal ligations in hospitals at the moment of delivery. To medically justify this procedure, which could be performed only in case of serious medical risks for the health of the mother, they performed it via caesarean section, leading to an increase in (often unnecessary) surgical deliveries. Compounding matters was the lack of public family planning services providing free contraceptives and of the international debt crisis that hit Brazil and more broadly Latin America in the 1980s, deepening social, gender, and racial inequalities. In a country where abortion was (and still is) illegal, except in the case of rape, maternal life risk and severe fetal brain injury, low-income women opting for a tubal ligation was a sort of constrained choice.

Despite the profound shift in the past two decades from population control via international interventions aimed at reducing fertility rates to promoting reproductive and sexual health and rights, the old neo-Malthusian rationales are still alive and well. As De Zordo convincingly shows, those neo-Malthusian biases coexist in Brazilian family planning clinics alongside the current and dominant discourse on reproductive rights, gender equality, and citizenship. She exposes the ensemble of discourses and medical practices that discipline low-income women, targeting their sexual, contraceptive, and reproductive life and stigmatize them when they deviate from neo-Malthusian rationales. The patronizing and dismissive attitudes of medical experts toward their clients starkly reminds us of the stubbornness of those Malthusian strains even as people and practices change. Physicians and nurses working in public family planning services in Salvador da Bahia (Brazil) do not think anymore that sterilization is the best solution to their patients' economic and family problems, such as poverty, criminality, and gender inequalities, glossed as '*machismo*'. Rather, they place hope in education – above all related to family planning. Every citizen, they state, has the 'right' to have access to good education and free family planning services, as the Brazilian Constitution established. Health professionals' mission, therefore, is no longer to persuade poor women to limit births and eventually have tubal ligations, but to 'enlighten' them with the biomedical 'instrumental rationality' (Good 2003). Most health professionals interviewed by De Zordo believed that, once 'enlightened' by the bio-medical knowledge, their low-income, 'ignorant', female patients would automatically abandon the 'absurd' beliefs, 'myths' (their 'culture') and ineffective (non-medical) contraceptives they still use and adopt medical, temporary contraceptives.

Tubal ligation, prized in the past as the most 'rational' course for poor women, is now labeled by health professionals working in family planning services as 'irrational'. As De Zordo shows, different biopolitical rationalities – the old neo-Malthusian rationale and the new rationale of sexual and reproductive rights – coalesce, producing a double stigmatization of female family planning users as 'victims' (of social and gender inequalities) and 'irrational' patients, 'irresponsible' mothers and 'bad' citizens if they do not embody the neo-Malthusian and biomedical rationales shaping medical practice.

Concerns with disciplining reproduction in the Islamic context of Pakistan, as Varley observes, take an interesting twist as historically secular family planners deploy moderate interpretations of Islam to promote a 'small family' model. The idea is to disseminate the value of a small family as a healthy and economically sensible norm. Aiming to get women and men to conform to such a norm and foment within them a desire to aspire to that norm exposes the essence of biopower.

This shift in biopolitical rationalities from secular to 'Islamized' family planning integrates moderate Islamic juridical and religious norms concerning sexuality and contraception. In Gilgit-Baltistan, where Varley undertook her fieldwork, this process of 'Islamization' has been aimed at making family planning programs more effective and at involving in particular Sunni women, who are a minority in this specific region and whom health professionals label as 'conservative', 'backward', and resistant to family planning. Sunni women's bodies and fertility have therefore become the battlefield of what Varley defines as 'Islamic biopolitics', which has fuelled existing ethnic and religious tensions and put Sunni women in a difficult position.

Varley compellingly explores, in her paper, the double rationale inspiring the new 'Islamized' family planning programs carried out: On the one hand, the neo-Malthusian rationale still justifies family planning aimed at reducing fertility rates, highlighting the positive impact that limiting births has on women's, children's and families' health and economic well-being; on the other hand, the moderate religious, Islamic rationale emphasizes the positive impact of contraceptive use in married couples, for the woman's and her children's well being but only if aimed at spacing births.

In a region marked by recent violent clashes between Sunni and Shia groups that have fuelled Sunni conservative *ulema*'s (Muslim clergy) pronatalist campaigns, the coalescence of these two biopolitical rationalities, Varley argues, leads to the stigmatization of both Sunni women who have large families and of Sunni women who have or wish to have small families. The first ones are in fact labeled as 'irrational' by health professionals working at family planning services; the second ones are stigmatized by their own families, communities, and conservative *ulema* as immoral Muslims.

As the papers collected in this special issue show, along with the neo-Malthusian, economic, and biomedical rationalities, religion also strongly influences reproductive policies and behaviors both in the 'global North' and in the 'global South'. In Catholic countries such as Italy, politicians and clergy defend the family and the rights of the embryo. As a result, Marchesi argues, Italian women are accused of being 'irrational' and 'irresponsible' if they ask to access assisted reproductive technologies, use sperm donors, and implant fewer embryos than those produced during the in-vitro fertilization cycles, threatening these embryos' 'lives'. In Poland, women cannot legally interrupt an unintended pregnancy and are morally condemned by the Catholic Church if they do not have large families.

Enticements to have large families come up against other structural factors: discrimination that many women encounter in the job market, the inability of women to count on a strong Welfare State or on policies supporting the reconciliation of family and work, and gendered divisions of labor in which many women struggle with deep gender inequalities in the division of domestic work as well as children's and elder care.

In these contexts, what constitutes 'rational', or 'responsible' reproduction? The answer to this question, the authors of this special issue brilliantly show, radically changes depending on who is asked and what perspective is taken. Within Europe, demographers stigmatize the 'excessive' fertility of immigrant women who are expected to embody a more 'rational' and responsible reproductive behavior by limiting births and controlling their fertility. At the same time, demographers create alarmism around the low fertility of European countries. This 'demographic alarmism' (Krause 2006, quoted by Marchesi, this issue) legitimizes and increases xenophobic fears concerning immigrants' reproduction in countries where the vast majority of migrant women move from the 'global South' and the post-socialist countries. The flux of domestic workers, nannies, and adoptees literally support the reproduction of European families. They are seen as needed to compensate the decrease in fertility rates on the one hand and the dismantling of the Welfare State and deep gender inequalities on the other.

In Italy as well as in Poland, children have become a 'luxury'. Most Polish Catholic women interviewed by Mishtal had decided in fact to postpone motherhood or to have only one child, even against their desire to have more children, because they felt responsible towards their families and feared losing their jobs. They also wished to be good mothers, able to take care of their children. Their 'decision' to postpone motherhood or to have one child emerges from their accounts as the opposite of a selfish act.

Marchesi questions the 'rationale' of European gynecologists who label as 'irrational' women's 'choice' to prioritize the search for a job and therefore postpone motherhood until it is biologically difficult to have children. In a context of economic crisis and unstable jobs, she argues, this is actually the most 'rational' decision women feel they can make. In Brazil, on the contrary, experts label as 'irrational' low-income women who do not effectively use temporary contraceptives to postpone motherhood and rationally plan small families.

To put the case studies into perspective, Morgan and Roberts recast Foucault's framework as 'reproductive governance', a move that provokes an examination of the links among moral regimes that are deeply embodied, the national political strategies that lend them definition, and the global economic logics that underwrite them. Pushing a 'politics of life' perspective (after Fassin 2007), Morgan and Roberts draw attention to the ways in which discourses, policies, and protests related to reproductive and sexual behaviors, boldly framed as human rights, have come to be used and abused in unpredictable ways. Indeed, the morphing of 'rights' discourse from one of reproductive freedom for living and breathing humans to one of divine rights for future imagined humans is unanticipated.

Negotiations

A number of the papers demonstrate the limits of new biopolitical regimes, how women and advocates speak back to them, how they negotiate the unjust dimensions, and how they struggle to reconcile discrimination, instability, care, and work – or lack thereof. The papers expose commonalities that reveal the agency of those who are the targets of reproductive policies: (1) a diverse yet conscious resistance against and awareness of efforts to manage life and discipline sexuality; (2) a strong yet unpredictable entanglement with the pernicious role of rationality in efforts to reinforce biopolitical projects.

In Brazil, activists of the feminist and black movements denounced family planning campaigns and practices as eugenic programs aimed at limiting birth among poor black people. Research supported their claims, finding that women with lower educational levels and economic statuses underwent tubal ligation at a younger age compared with their upper-middle class peers. Given that blacks in Brazil tend to be overrepresented among the poorer classes, clearly these practices impacted black women at a higher rate than those of the general white population. A heated political and scientific debate ensued. Feminists and black activists pressured politicians, who in 1996 passed legislation to reform family planning and offer a number of free reproductive health services. The result has been a decrease in female sterilization rates (to 29% by 2006), a slight increase in male sterilization (to 5%) and an increase in other contraceptives such as condoms, hormonal injections, and the IUD.

As family planning in Brazil has gradually shifted toward contraceptives, De Zordo lucidly documents how low-income participants in family planning courses reacted to the experts who viewed them as 'ignorant' as they aim to 'enlighten' them. There, the gaps between scientific and folk understandings of hormones and contraceptives generated misunderstanding, anxiety, and mistrust. Doctors and medical personnel tended to ignore patients' complaints; meanwhile, young women were capable of clearly expressing experiences and fears of weight gain, bloating, headaches, and nervousness. Furthermore, the participants spoke about the challenges they faced in following mathematical regimens of pill taking. Some noted long and irregular work schedules. Others lacked health insurance and access to expensive contraceptives. Still others spoke about the difficulty of planning in the heat of the moment and the risk involved in saying 'no' to sex or insisting one's husband use a condom. The fact that experts habitually dismissed these perspectives and seemed unaware of structural constraints, and yet that the 'patients' continued to assert these limitations – whether in classes, in waiting rooms, or in conversations with anthropologists – exemplifies the ways in which subjects of biopower find themselves entangled with rationalist biopolitics in an ongoing, intimate, everyday way.

In Poland, some feminist groups during the past two decades have been persistent in condemning the government's neoliberal agenda for the negative impact on women's lives and reproductive decisions and practices. These proclamations are often marginalized or, when State and Church officials do tune in, they often dismiss these claims as anti-family and communist. In both Poland and Italy, feminist groups usually propose effective pro-family programs, i.e., more public nurseries and part-time jobs, longer maternity and paternity leaves, than the political parties influenced by the Catholic Church supporting pronatalist campaigns.

Definitions of what constitutes 'responsible' reproduction collide in the context of 'replacement anxieties', as Marchesi demonstrates with the Italian case, where a cacophony of voices can be heard. This, she notes, 'is far from a settled issue'. Feminists cast responsibility in terms of self-determination and note the challenges of having and caring for a child as neoliberal restructuring of the economy has delivered a precarious labor market; politicians' framings add a decidedly disciplinary and nationalistic flair; social service workers scrutinize and manage reproductive practices of migrant women; the Church chimes in and adds a regulatory chord that echoes Vatican notions of contraception, sexuality, and the family; and medical experts send messages about protecting the womb as they emphasize safe sex and declining fertility over the life course.

Far from Europe, in Northern Pakistan, Sunni women struggle to defend their own rationale against on the one hand neo-Malthusian family planning programs and on the other hand local conservative *ulema*, who invite them to have as many male children as they can if they want to behave as good Muslims and contribute to nationalistic struggles. Some who live in expensive urban areas where they lack the support of large families prefer having small families and therefore use contraceptives. Others, however, mainly in small villages, where they have their large families' support, desire to have many children and therefore do not regularly use contraceptives. Social pressure from family members (mainly mothers-in-law) and local clergy also changes from smaller to bigger cities; therefore, the meaning of making a 'rational' reproductive choice varies from one context and one woman to the other.

From Asia to Europe and Latin America, religion is able to produce and reproduce strong and influential 'moral regimes', but these do not go uncontested. In fact, as Morgan and Roberts remind, Latin American 'Catholic women ... account for some of the highest abortion rates in the world'. Moreover, despite the Catholic Church's condemnation, assisted reproductive technologies flourish in many Latin American countries. And even in a European country such as Italy, where the Catholic Church wields its influence on politics and reproductive health, not only has abortion been legalized since 1978, but – as Marchesi points out – the 2004 restrictive law on medically assisted reproduction has since been contested and partially liberalized.

Hence, in response to the proliferation of bio-political '(ir)rationalities' new and creative forms of contestation emerge. Polish women refuse in practice, Mishtal argues, to embrace the role of 'responsible biological citizens' and to reproduce the ideal Polish, Catholic family, by limiting their births. However, they have difficulties in producing effective movements of contestation, able to actually challenge and change national policies. Italian feminist groups, in contrast, creatively and with varying degrees of success contest on the ground the nationalistic, xenophobic, and pronatalist campaigns and the Catholic defense of the rights of the embryos pitted against those of women.

Biopower produces truth and moral regimes that shape reproductive policies at national and international levels. Access to safe and legal abortions, therefore, varies across the globe but, as Morgan and Roberts show, such access reveals unpredictable patterns, with a spectrum that ranges from liberal to restrictive reproductive policies. Whereas Mexico City and Colombia have liberalized abortion, other countries, such as the Dominican Republic, El Salvador, and Nicaragua, have totally banned it. Moreover, in the same country – for instance, in Brazil – restrictive laws on abortion and liberal laws on medically assisted reproduction coexist. Furthermore, conservative governments have passed progressive laws and vice versa.

Shifts in the rationality of reproductive governance always produce new moral regimes, Morgan and Roberts argue. Today, the struggles around reproductive policies are articulated in juridical terms much more than in the past, and produce rights-bearing citizens pitted against each other – embryos versus women, native versus immigrant women and heterosexual versus homosexual women and couples. These new moral regimes generate social and political spaces for ongoing negotiation.

Implications

As a collection, these papers powerfully demonstrate how biopolitics do not simply flow downstream, as a river traceable to a single spring and then contained between two banks, but rather stem from manifold sources and generate multiple tributaries – what might be described as a proliferation of polyvocality. Inspired by Foucault's theories on biopolitics and biopower and by a long tradition of feminist anthropological studies on reproduction, the authors of this special issue ultimately examine how reproduction is shaped in different geo-political contexts, from Europe to Latin America and Asia, and expose how different bio-political projects, strategies, and effects are globally related. Based on meticulous ethnographies, they detail how biopolitics operate in diverse social and cultural contexts and produce not only scientific and religious 'truth regimes' but also tangible 'moral regimes' (Morgan and Roberts, this issue). These regimes aim at regulating reproductive and also sexual practices as well as gender relations. Furthermore, they engender 'new kinds of... groups and individuals, who increasingly define their citizenship in terms of their rights (and obligations) to life, health and cure' (Rabinow and Rose 2006, 203), including entities such as non-governmental organizations (NGOs), social movements, and other interest groups. Morgan and Roberts also emphasize that the emergence of new moral regimes seems to be bringing about a shift in 'population', as a concept, as reproduction and life itself come to be conceptualized anew. The so-called new rationality of 'reproductive governance' extends individual rights differentially. Immigrants are seen as 'resource depletors' and hence can be denied rights, and this group is now pitted against not only bona fide citizens but fetus-as-citizens. The litmus test appears to come down to who takes what from the neoliberal State, according to Morgan and Roberts.

In the final analysis, it becomes clear that rationality runs rampant as an assumption that experts deploy when it comes to explaining human behavior. Elites in Brazil assume that poor people lack rationality along with responsibility. Drawing on demographers' projection of a need for 'more coffins than cribs', Polish politicians and pundits called for taking on 'the battle in the bedroom', demonizing women for their concern over jobs and ignoring the hardships that economic restructuring has delivered. Politicians in Italy assume that as Italians have so few babies they are reproducing nothing other than a dying population, suggestive of rationality unhinged. In the latter case, Marchesi exposes rationality and its slippages, in the complex interplay of demographic anxieties, Catholic influences, reproductive technologies, and neoliberal reforms that she reveals with vivid attention to emergent subjectivities. Governance, whether North or South, masquerades as being above the pale of irrationality and is often dressed in the attire of bourgeois nationalism.

Indeed, bio-medicine plays a key role in producing and reproducing bio-political regimes of truth that socially and morally justify the management and administration of the reproduction of specific populations and groups of individuals, whether at national or transnational levels. These papers show how nation-states and their laws shape the authority of bio-medicine – more liberal in some contexts than in others – and how different subjects contest truth regimes on the ground. As Rabinow and Rose point out, nowadays the power of bio-medicine and of its agents 'to let die' at the end of life, the start of life or in reproduction, are simultaneously enhanced by medical technology and regulated by other authorities as never before' (Rabinow and Rose 2006, 203).

At the same time, this collection outlines how the progressive privatization of healthcare and the dismantling of Welfare States resulting from the implementation of neoliberal policies is increasing social and gender inequalities that directly affect reproductive practices and decisions on the ground. This transversal process is occurring both in the 'global North' and in the 'global South'. In this neoliberal context, reproductive policies often contribute to reinforce and deepen existing inequalities, by targeting some individuals and groups, such as the poor, and/or immigrants, with programs aimed at controlling and limiting their fertility, while giving others – a privileged minority – the possibility of having children, sometimes despite their infertility.

These papers therefore show something that 'is almost absent in Foucault's work', as Fassin notes: 'the fact that biopolitics has consequences in terms of inequalities' (Fassin 2009, 53). They show in particular that although biopolitics concerns population it is more importantly 'about life and more specifically about inequalities in life which we could call bio-inequalities' (Fassin 2009, 49). In this sense, biopolitics has a normalizing effect on how people may live their lives.

Ultimately, when we put the voices side by side, assumptions are revealed. Projects delineated. A sustained and penetrating gaze at the ways in which so-called experts mobilize rationality and 'culture' across contexts expose fissures in biopolitical agendas. The presumed rationality that underwrites demographic governance crumbles. Empiricist fallacies are unveiled. Drawing out and exposing these fissures may be the most compelling resource for carrying out an engaged and empathetic genealogical research agenda that re-values subjugated forms of knowledge and exposes why such endeavors matter.

Acknowledgements

The papers in this special issue are based on a panel entitled, 'Irrational reproduction: population politics and practices at the intersections of North and South', presented at the AAA (American Anthropological Association) Annual Conference, San Francisco (USA) on 22nd November 2008. All authors presented a paper with the exception of Emma Varley and Morgan-Roberts, who joined our special issue later on.

References

Agamben, Giorgio. 1998. *Homo sacer: Sovereign power and naked life*. Stanford, CA: Stanford University Press.

Coale, Ansley J., and Susan Cotts Watkins, eds. 1986. *The decline of fertility in Europe*. Princeton, NJ: Princeton University Press.

Bourdieu, P. 1987. The force of law: Toward a sociology of the juridical field. *Hastings Law Journal* 38, no. 5: 814–53.

Ehrlich, Paul R. 1968. *The population bomb*. New York: Ballantine Books.

Fassin, Didier. 2007. Humanitarianism as a politics of life. *Public Culture* 19, no. 3: 499–520.

Fassin, Didier. 2010. Coming back to life: An anthropological reassessment of biopolitics and governmentality. In *Governmentality: Current issues and future challenges*, eds. Ulrich Bröckling, Susanne Krasmann and Thomas. Lemke, 185–200. New York: Routledge.

Fassin, Didier. 2009. Another politics of life is possible. *Theory, Culture & Society* 26, no. 5: 44–60.

Foucault, Michel. 1990[1978]. *The history of sexuality: An introduction, Volume I*. New York: Vintage Books.

Foucault, Michel. 1980. 'Two lectures'. In *Power/knowledge: Selected interviews and other writings 1972–1977, Michel Foucault*, ed. Colin Gordon, 78–108. New York: Pantheon Books.

Foucault, Michel. 2004. *Security, territory, population: Lectures at the College de France 1977–1978*. New York: Palgrave.

Fukuyama, Francis. 1992. *The end of history and the last man*. New York: The Free Press.

Ginsburg, Faye, and Rayna Rapp. 1995. *Conceiving the New World Order: The global politics of reproduction*. Berkeley, CA: University of California Press.

Good, Byron. 2003. *Medicine, rationality and experience*. Cambridge: Cambridge University Press.

Greenhalgh, Susan. 1995. Anthropology theorizes reproduction: Integrating practice, political economic, and feminist perspectives. In *Situating fertility*, ed. Susan Greenhalgh, 3–28. Cambridge: Cambridge University Press.

Greenhalgh, Susan. 2003. Planned births, unplanned persons: 'Population' in the making of Chinese modernity. *American Ethnologist* 30: 196–215.

Greenhalgh, Susan, and Edwin Winckler. 2005. *Governing China's population: From Leninist to neoliberal biopolitics*. Stanford, CA: Stanford University Press.

Gal, Susan, and Gail Kligman. 2000. *The politics of gender after socialism: A comparative-historical essay*. Princeton, NJ: Princeton University Press.

Hacking, Ian. 1990. *The taming of chance*. Cambridge: Cambridge University Press.

Horn, David G. 1994. *Social bodies: Science, reproduction, and Italian modernity*. Princeton, NJ: Princeton University Press.

Johnson-Hanks, J. 2008. Demographic transitions and modernity. *Annual Review of Anthropology* 37: 301–15.

Krause, Elizabeth. 2005. *A crisis of births: Population politics and family-making in Italy*. Belmont, CA: Thomson/Wadsworth.

Krause, Elizabeth. 2006. Dangerous demographies and the scientific manufacture of fear. *The Corner House*, Briefing paper, 36. http://www.thecornerhouse.org.uk/

Lock, Michelle, and Judith Farquhar. 2007. *Beyond the body proper*. Durham, NC: Duke University Press.

Malthus, Thomas. 1798. *An essay on the principle of population*. London: J. Johnson in St. Paul's Church-Yard. http://www.esp.org

Martin, Emily. 2001 [1987]. *The woman in the body: A cultural analysis of reproduction*. Boston, MA: Beacon Press.

Marx, Karl. 1906 [1867]. *Capital*. New York: International Publishers.

Oksala, Johanna. 2007. *How to read Foucault*. New York: W.W. Norton.

Paxson, Heather. 2004. *Making modern mothers: Ethics and family planning in urban Greece*. Berkeley, CA: University of California Press.

Rabinow, Paul, ed. 1984. *The Foucault reader*. New York: Pantheon Books.

Rabinow, Paul, and Rose Nikolas. 2006. Biopower today. *BioSocieties* 1: 195–217.

Rapp, Rayna. 2000. *Testing women, testing the fetus: The social impact of amniocentesis in America*. New York: Routledge.

Rose, Nikolas. 1996. *Inventing our selves*. Cambridge: Cambridge University Press.

Rose, Nikolas. 2007. *The politics of life itself: Biomedicine, power, and subjectivity in the twenty-first century*. Princeton, NJ: Princeton University Press.

Schneider, Jane C., and Peter T. Schneider. 1996. *Festival of the poor: Fertility decline and the ideology of class in Sicily: 1860–1980*. Tucson, AZ: University of Arizona Press.

Willis, Paul, and Mats Trondman. 2000. Manifesto for ethnography. *Ethnography* 1, no. 1: 5–16.

Irrational non-reproduction? The 'dying nation' and the postsocialist logics of declining motherhood in Poland

Joanna Mishtal

Department of Anthropology, University of Central Florida, Howard Phillips Hall 309, Orlando 32816, USA

Polish birthrates during the state socialist period, 1948–1989, stayed above replacement level but since 1989 fell dramatically to one of the lowest in Europe, at 1.29 in 2010. The Polish Catholic Church and the newly-elected nationalist government of Lech Wałęsa reacted by escalating pronatalist rhetoric calling on women to increase childbearing in the name of economic and nationalist causes. Reflecting the renewed dominance of the Church, Wałęsa implemented restrictions on family planning, including abortion, contraception, and sex education, justifying them in moral and demographic terms. Plummeting fertility has been portrayed by the Church, media, and state as dangerous and unreasonable – a sign of Polish women's rejection of motherhood and the embrace of selfish priorities. Simultaneously however, the state cut back motherhood-friendly policies established by the socialist regime, including subsidized childcare, maternity leave, and healthcare. This paper draws on 19 months of fieldwork between 2000 and 2007, using interviews with 55 women in four healthcare clinics in Gdańsk area, and participant-observation at the social services offices in Krakow. This paper shows that far from irrational rejection of motherhood, Polish middle-class women are guided by pragmatic reasons when delaying parenthood in order to navigate the new political landscape marked by job insecurity and gendered discrimination in employment. Yet, rather than implementing work-family reconciliation policies that have stimulated fertility elsewhere in Europe, the Church and state insist on blaming women for 'irrational' non-reproduction, thus betraying a lack of political commitment to gender equity in employment, reproductive health, and in the family.

Introduction

In November 2002, when Poland hit the bottom of its fertility drop with 1.22 children per woman, the major political analysis weekly, *Polityka*, announced 'There are too few of us' ['*Malo nas*'], warning that the Polish family is dying out due to a shift of young women's priorities – a shift that goes 'against the natural characteristics of the female psyche and degenerates the economic prospects of the nation' (*Polityka* 2002:13). These sentiments express a generalized anxiety about the decline of the heterosexual Catholic family perceived to be threatened by low fertility. That same year, the First Polish Demographic conference held in Warsaw for the purpose of assessing the results of the recent population data, dramatically declared: 'This year we'll need more coffins than baby cribs'. Such warnings led the press to this rhetorical question: 'Shouldn't Polish women be thinking more about their first child, instead of their first job?' (Henzler 2002). As the demographic 'crisis' continued in subsequent years with negligible signs of improvement, *Gazeta Wyborcza*, the largest Polish daily newspaper suggested again that we look to women and their inability or lack of desire to birth, and proposed that for 'Europe: looking to regain its lost sense' it must take on 'the battle in the bedroom', i.e., 'If women in the EU don't start birthing more, we won't be able to reverse the escalating trend of aging in Europe' (Pawlicki 2005). Common imagery in the media and political discourses depicts Poland as 'depopulated' – indeed, the nation has experienced negative population growth for the 20 years since state socialism collapsed. For a nation with 89% affiliation with Roman Catholicism and restricted family planning options, including a ban on abortion, a major fertility decline might seem surprising.

Fertility decline is not new. In the preceding few decades a decline has been observed in most of the world. However, the situation in Eastern Europe is particularly dramatic because fertility plummeted across this region at a rapid rate only after the Soviet Union's collapse. During the years preceding the present slump, Poland's total fertility rate (TFR) had been at or above replacement of 2.1 since the Second World War. After 1989, Poland's TFR plummeted to 1.22 by 2002[1] and increased marginally to 1.29 in 2010 – the third lowest TFR in the EU (after Czech Republic and Lithuania).[2]

Demographers studying declining fertility around the world took a keen interest in the Eastern European phenomenon, and began to describe trends in Poland that were seen elsewhere: increasing mean age at first childbearing, declining marriages, and rising extramarital births, divorces and cohabitations (Fratczak 2004). To explain this phenomenon, some scholars have proposed the 'tempo effect', arguing optimistically that the very low fertility in Eastern Europe is driven by the postponement of childbearing, and that the system will eventually 'catch up' (since it is assumed that women cannot postpone births indefinitely due to biological barriers), following which fertility will at least partially recover, making the current situation just a temporary adjustment to new economic circumstances (Bongaarts 2002).

As fertility in Eastern Europe showed no signs of recovery during the last 20 years, demographers began to look to social and cultural factors. A recent and more convincing argument put forth by Reher (2007) and McDonald (2006) presents an opposing view to the 'tempo effect'. They assert that while the postponement of childbearing is an important factor in European fertility decline, 'extremely low fertility has been around for too long for it to portend anything other than major long-term social change' (Reher 2007, 194). Significantly, new demographic theories propose that gender inequities experienced by women are decisive in discouraging motherhood as the burden of carework combined with desired, or necessary, employment proves too overwhelming for many women to manage (McDonald 2000).

Yet, the persistent myth of a temporary phenomenon and the concomitant disregard for the gender inequities theory in fertility decline have prevented many European states from taking policy action to make motherhood more feasible. The Polish state responded in 1999 with the 'Profamily Program', developed based on the Vatican's Family Rights Charter.[3] The Program consists mainly of political rhetoric that calls for higher childbearing and stresses the importance of a nuclear family, but is void of any meaningful policies to help women combine work and motherhood. The Program succeeded in making expressions such as 'demographic low' (*niż demograficzny*), pronatalist politics (*polityka pronatalistyczna*), and profamily politics (*polityka prorodzinna*) a quotidian part of the Polish vernacular. But the Program has utterly failed in its mission as birthrates have fallen even further since the Program's launch in 1999.[4]

Without work–family reconciliation policies on the horizon, the experience of very low fertility in Poland has now become a structural problem with virtually certain long-term consequences. The complexities of the Eastern European demographic situation and the culturally-specific differences between the nations in this region call for caution in applying unidirectional transition theories that imply a predictable evolutionary process of social or economic change (Rivkin-Fish 2003). New fertility research in Poland requires attention to the institutional and cultural context – in particular the role of the Catholic Church on national policy-making level, the conservative state administrations, and the nationalist discourses – within which women negotiate their new circumstances, and must explore women's experiences and perspectives on motherhood before the observed macro population trends can be theorized and understood.

Attention to the role of the institutions of the Church and state in East European fertility decline builds on philosopher's Michel Foucault's concept of biopolitics. Foucault argued that the state began to manage individuals beginning in the eighteenth century through a variety of disciplining and surveilling mechanisms, while simultaneously the state set out to govern the larger populations by using a systematic census-like accounting of behaviors. In the context of fertility decline, demographic knowledge and pronatalist policies constitute the expression of what Foucault refers to as biopower – a set of state techniques that 'brought life and its mechanisms into the realm of explicit calculations and made knowledge-power an agent of transformation of human life' (Foucault 1980, 143). He notes a consolidation of state power via new administrative attention to the health of populations, but reconfigured in the service of the state. In Poland, the notion of population health is recast as numerical population growth, and women are simultaneously recast as 'biological citizens' expected to live responsibly and rationally in order to maximize their contribution to the growth of the population (Rose and Novas 2005). The statistical surveillance of women's reproductive behaviors that generate annual total fertility rates and birthrates and the accompanying demographic knowledge serve at the state's disposal to substantiate the implementation of state policies aiming to reverse negative population trends. In other words, biopower is simultaneously both the production of knowledge or 'truth discourses' about, in this case, reproductive behaviors, and the response to these changes with particular policies aiming to 'improve' the life or health of the population (Rabinow and Rose 2006, 197). In the Polish case, repressive policies that are justified by the 'demographic crisis' are restrictions on access to family planning and sex education.

Given the aggressive political and media rhetoric in Poland raising alarm about a demographic crisis and women's presumably confused priorities, several questions called for consideration in this research: how do Polish women explain their reproductive decisions in light of the state and Church's accusations of their irrationality? What concerns do they consider central to their reproductive strategies? If the majority is Catholic, how can women justify going against the Catholic principles that disallow deliberate fertility control?

In light of these questions, the research in Poland focused on the postsocialist logics of motherhood: the ways in which women understand their life choices and navigate the postsocialist web of constraints in family planning, employment, care work, and culturally and religiously specific notions of 'proper' motherhood and 'proper' number of children. In particular, women who work in the middle and upper socioeconomic class settings have experienced both types of pressures from their coworkers: to have a child, but to not have more than one child – this is emerging as an important class marker of middle-classness. Thus, this study contextualizes women's experiences within the historical shifts that Poland experienced as it moved from state socialism to independence, with politico-economic, juridical and cultural implications. Arguments here are also framed using recent Vatican's statements regarding the demographic situation in Europe and women's role in it.

This paper draws on 19-month fieldwork in Poland between 2000 and 2007, especially focusing on research in 2007 in Gdańsk where interviews with 55 women aged 18–40 were conducted in four healthcare clinics. Secondarily, it draws on participant-observation in the social services offices in Krakow. For confidentiality, all names used in this paper are pseudonyms. On average, the women in the Gdańsk research had 0.92 children, which approximates Poland's urban TFR of 1.1, and 94% affiliated with Catholicism – a somewhat higher percentage than 89% nationally. Overall, the study sample was more educated than the national average, which reflects the predominantly middle socio-economic class status as well as the urban nature of the sample.

The makings of a demographic crisis

To understand the historical context for the post-communist drop in fertility it is vital to consider the complexity of changes that women experienced in a shift from the old state socialist regime to the new independent Poland after the Soviet collapse in 1989. Fertility fell across the entire geopolitical region, positioning Eastern European nations in the majority of so-called very low fertility countries with TFR below 1.5. Despite the region's heterogeneity, the shared legacy of state socialism and similar global economic pressures result in many common post-1989 experiences that are significantly associated with fertility decline.

Before 1989

Soviet controlled state socialism was established in Poland in 1948. Marxist egalitarian principles drove many new policies that greatly expanded education and employment, especially for women, and offered free healthcare. State socialism created an inflexible employment structure that provided a great deal of work security with virtually no threat of unemployment. This was significant for women who were able to enter the job market, interrupt work for childbearing, and return to the safety of their jobs with minimal or no loss of wages. Consequently, Polish women's full-time employment rose to 78% during state socialism (Fodor et al. 2002, 371–2).

Equally significantly the communist governments across Eastern Europe expanded access to reproductive rights (except in Romania). In 1952, the Polish state legalized abortion, began to subsidize contraception, and introduced sex education in schools, and prescription contraceptives began to be subsidized at 70% (Mazur 1981, 196). The Church strongly opposed the legalization of abortion, and the increased access to family planning; however, the Church's power to influence policy was tenuous at best, given the strong secular nature of the communist regime. This separation of Church and state indeed benefited women.

Although the regime was unable to reduce gender inequalities within the family – women were still expected to manage the household and care for children, husbands, and relatives – gender relations were certainly reconfigured as women pursued careers, financial independence, and greater reproductive autonomy.

After 1989

Following the regimes' collapse, Eastern European nations underwent profound economic transformations, shifting from the security of welfare states to the instability of free market economies. The policies that had been critical in helping women to reconcile work and family were rapidly dismantled. In Poland, the situation was particularly ironic since the regime's fall was hastened by the opposition waged by Solidarity – the Catholic-nationalist labor union. As the union's leaders, under Lech Wałęsa, took power after 1989, the new government wholeheartedly embraced neoliberal economics giving market forces primacy in solving economic and social problems to the detriment of workers and organized labor. President Wałęsa, himself a former electrician in the Gdańsk shipyard, rapidly proceeded with anti-worker reforms that targeted women, including major cuts in maternity leave, childcare and other social services.

Comparative studies of welfare provisions in Eastern European nations show that Poland has had one of the harshest reductions in family and maternity benefits (Fodor et al. 2002, 477–83). Wałęsa also launched healthcare privatization by, for example, cutting medicine subsidies from 100% to 35%, the lowest in the European Union (*The Economist* 2004).

In an effort to repay the Catholic Church for supporting the Solidarity movement against the socialist regime, Wałęsa immediately sought to ban abortion: by 1991 the Conscience Clause policy allowed individual physicians to refuse reproductive healthcare, in particular abortion (still legal at this time), and contraceptives by invoking conscience-based objections. This policy has since been abused by being invoked systemically by entire hospitals, rather than by individual doctors, and served as the stepping stone to ban abortion (Mishtal 2009a). Opponents of the looming abortion ban, in particular the Federation for Women and Family Planning, who were aware of the national polls favoring legal abortion, campaigned for a national referendum. However, a referendum was never held and the abortion ban became law in 1993. The new law criminalizes abortion under all but three circumstances: when the woman's life or health is in danger; when a prenatal test shows a serious fetal deformity; or the pregnancy resulted from a crime. Since 1956, only 3% of all abortions had been performed for these reasons, thus, the ban de facto excludes 97% of previously performed abortions (Johannisson and Kovcs 1996; Nowakowska and Korzeniowska 2000). Within the EU, Poland's law is second in severity only to Ireland's. Access to contraceptives has also been reduced as the state eliminated all subsidies from the health insurance coverage by 2002. Simultaneously with the reproductive rights restrictions, the Church and state intensified the traditional gender rhetoric emphasizing the primacy of motherhood. Women's rights were routinely portrayed as discouraging motherhood, and precluding the well-being of the family and 'normal' biologically destined gender roles. Feminist groups' counter-discourse contesting the Church's campaign was dismissed in the right-leaning politics as communist.

In the area of employment, job security was no longer guaranteed, unemployment soared, especially among women, and preferential hiring practices began to favor applicants with inside connections. The World Bank's assessment of Eastern Europe observed that 'the transition from planned to market economy has witnessed one of the biggest and fastest increases in inequality ever recorded'.[5] Indeed, economic changes are driving a new socioeconomic class stratification in Poland – the Polish Gini coefficient (a measure of the inequality of wealth distribution between 1 (complete inequality) and 0 (complete equality)) went from 0.26 in the 1980s to 0.36 in 2005. Currently, Poland's Gini coefficient is second highest after Russia in the post-Soviet region.[6] Polish women are bearing the brunt of this stratification; they have twice the likelihood of falling below the poverty line as men and constitute the majority of the unemployed due to acute gendered discrimination in the workplace (Domanski 2002, 393). Feminist groups are highlighting the detrimental effects of neoliberal economics but are often dismissed by the Church and state as communist and anti-family. It is in a response to this historical, politico-economic, and ideological shift experienced in the political, legislative, and public realm that Polish women began to postpone motherhood.

The postsocialist logics of declining motherhood

'So I'm happy we're having a demographic crisis, it's what the government deserves!' exclaimed Alina, a 40-year-old information technician who, at the time, had an 18-year-old son. She explained that the core of the issue lies in the dwindling social support that her family enjoyed before 1989. She went on: 'I'm so mad at the government, I think it's great that we're having a demographic crisis because the state doesn't give us any support – no support for women who are pregnant or women with kids. I had my son just before 1989 and it was no problem to have kids back then, even though I was a single mother' (interview with Alina: Gdańsk 2007). Sentiments like these, expressing both anger squarely directed at the postsocialist policies and almost nostalgia for the communist era when women were able to reconcile work and family, dominated many of the conversations during this fieldwork.

Confirming anthropological findings in Eastern Europe, women in this study who grew up during state socialism perceived motherhood and work as simultaneous endeavors, an aspiration that is proving far more difficult to attain after the fall of the regime (Erikson 2005). Combining work and family is implied in the pragmatic sense of 'self-realization' [*samo-realizacja*] that is part of Polish women's vernacular, if in particular among the middle-classes, and the imagined sequence of life events reflects this thinking, as Mariola, a 21-year-old student at the Jaggielonian University in Krakow explained:

> Really, women don't think anymore that they should give birth *first*, and then, later, they'll get everything else figured out. Now we think, 'Finish the degree, find some kind of work, in the meanwhile it might be good to find a husband', and then at that point, after some time, think of having a child. Among my friends the dominant opinion is that a child is a luxury. If the material situation is normalized enough, then that couple, or that woman, can decide to have a child. Still, the majority of women want at least that *one* child. (Interview: Krakow 2002)

Yet, postponement of pregnancy in favor of securing jobs has been depicted by the Church and the right-leaning media as a rejection of motherhood. Nearly all of the women in this research, regardless of education or employment, explained that far from rejecting motherhood, their desire is to have at least one but preferably two children. A common expression heard in the interviews was 'the first child is for oneself, the second child is for the first', which was explained to mean that parents need a child to take care of them in their old age, while the second child is important to prevent the first one from becoming selfish or spoiled. In reality however, women felt that a second child is rapidly becoming an economic luxury, especially because of lack of childcare. The state eliminated subsidized childcare and babysitters are still rarely used since the idea is quite new, the service is very expensive and 'you simply can't trust strangers with your kids', as one of the women warned.

For many the only way to maintain one's job is to recruit one's *babcia* – literally, a grandmother – referring to the tradition of having a mother or mother-in-law provide childcare for the grandchild. However, with the increased geographic mobility in Poland, and the search for jobs in other cities, fewer relatives are on hand to provide childcare. Women who had *babcias* considered themselves lucky. The issue of childcare was closely intertwined with the way that employers viewed female job applicants. The experiences of Celina and Ewa that follow are emblematic of those of nearly all other women in this study, whether these experiences were their own, their relatives', friends', or co-workers'. Celina, a 34-year-old border customs officer without children, expressed some sense of job security in her current government position but related the following experiences from recent years:

> Of course I went through my share of problems before I got this position; I had three unpleasant situations. When I was looking for a job one of the employers asked me if I'm planning to get pregnant, and if so, whether I had a *babcia* at home on hand to do my childcare; I said I didn't. Another employer told me that he requires a current pregnancy test as a condition for the employment. I totally disagreed with this practice and just left. And another employer told me he'd give me the position if I declared I won't get pregnant for 2 to 3 years. I left there too. The main reason [that I don't have kids]: no childcare! I have no kids but ideally I would like two. I've been married for 9 years. (Interview: Gdańsk 2007)

Since Celina has no children but is in a long-term relationship, it was significant to understand her perspective on the government's calls for more childbearing and whether or not this rhetoric mattered to her. When asked, 'Are you concerned about the demographic crisis we are hearing about or rather not?' She immediately replied:

> Absolutely not. I'm very opposed to the current government, especially to its right-wing policies. My views are really opposite, so naturally the government has no influence on my thinking or my life. If a new leftist government called for higher fertility *and* provided protection for women who are interested in having kids, then I might consider [having a child]. My mother is a good example of what happened here: she was a single mother with two kids ages 19 and 7 right during the '89 transition and she lost her job as soon as Solidarity came to power. They fired a single mother! What we need is legal protection for single mothers and women against being fired. It was also easier in the old system to get a place for your child in the infant care center or in a preschool. Now there's no space and you have to pay. (Interview: Gdańsk 2007)

Ewa, who is a 33-year-old physician in general practice with two children and has been working for a number of years in a stable position at a major health clinic, was one of a few women who were able to afford a babysitter:

> I would like a third child but we would need childcare, first and foremost, and second, we'd need a larger apartment. We badly need a *babcia*! We had a couple of nannies, babysitters, but that didn't go well – one of them burned my child with coffee! My husband helped with the childcare a little bit too. But all in all, it's very draining financially to pay for childcare. (Interview: Gdańsk 2007)

When the discussion turned to the state's alarmist calls to women to have more children and its ubiquitous rhetoric of pro-family politics, Ewa was asked whether the state's concern about the demographic situation in Poland mattered to her. She replied: 'I can't say that the state matters for me that way. That's because the state first and foremost needs to change its politics toward women, to offer childcare and protection at work for women. Currently, the state's political position is *anti*-family, not pro-family'. Despite her relatively comfortable economic position, Ewa echoed many other women's understanding of the problem, arguing that 'major changes in policies that would guarantee greater social services, childcare, healthcare, and the protection of women in the workplace' are needed.

Deeper conversation about work experiences revealed that widespread fear of discrimination by employers against pregnant women, new mothers, and women with small children drives women's decisions to postpone having a child (Mishtal 2009b). The majority of women in this research either directly experienced gendered discrimination in employment or knew of women who had. These narratives reported problematic employer practices that had an adverse influence on women's fertility decisions, such as firing women who returned to work following maternity leave, and requiring women to sign a contract pledging not to get pregnant for two to three years as a contingency for their hire. Women also reported other illegal practices such as asking female job applicants if they have small children and hiring permanent employees to replace women on maternity leave. Hanna, a 34-year-old sales clerk in a long-term relationship and with one child but with a desire to have two related the following experience, which echoes other women's concerns who, like Hanna, are likely to forgo the second pregnancy in favor of employment:

> We don't want any more kids under the current circumstances. One is enough. Thankfully, I was able to return to my job after our son was born but I had only 3 months off. I made an agreement with my boss that I'll only take 3 months. It was also better for me since I needed the money; I couldn't really afford to go on the child-rearing leave. But besides, I had a *babcia* at home to do the childcare so I was able

> to work again. I was certainly concerned about my work because some of my friends were fired once they became pregnant, so it was on my mind. About a year ago my boss fired an acquaintance of mine when she got pregnant. After a few months I decided to ask him very casually about that. He said he had the *right* to fire her. I didn't want to pursue it any further so I dropped it. (Interview: Gdańsk 2007)

It is against Polish labor law to fire a pregnant employee. Hanna told me she was aware that the boss had no right but was afraid to challenge him or appear confrontational, in case she 'catches his eye in a negative way' (*podpadnie*). However, many employees are often unfamiliar with their rights, and likewise, bosses are unfamiliar or unwilling to follow the law. Katarzyna Kurkiewicz, a labor lawyer who specializes in gender discrimination and works closely with a feminist organization – The Network of East-West Women in Gdańsk – corroborated the experiences of the women in this study, saying that 'these practices are increasingly common in Poland'. Kurkiewicz explained that even though it is unlawful for the employers to ask women about their pregnancy status, whether they have children, or to sign a no-pregnancy agreement, these practices have become customary, leaving women with little choice but to oblige. The employer cannot fire a woman who states that she is pregnant; hence harassment techniques are used to drive women out 'voluntarily' (Interview: Gdańsk 2007). Legal recourse for victims has been practically nonexistent due to the extraordinary delays and expenses that accompany legal procedures in Poland.

Many women reported that employers discriminated against applicants with small children, using the interview process to identify women who might miss work because of childcare. After experiencing work discrimination and limited maternity leave with the first child, women become discouraged from having another. As Zofia, a 27-year-old sales clerk, who was in the eighth month of her first pregnancy at the time of the interview and in a long-term relationship, explained:

> We are in a bad financial situation, so after this pregnancy we'll just have to wait and see. I would really like two kids because a single child *needs* a sibling, otherwise it'll turn out selfish. [. . .] I'm not sure if I'll even have my job when I go back, because when I got pregnant my boss started to make things difficult for me, all of a sudden he made everything an uphill battle for me – every time I needed to go to the doctor he would make a fuss or any other requests he always made it look like I'm asking for too much; he wanted me out. It just wasn't convenient for him to have any of the employees pregnant. My doctor took me off work when I was in my fourth month because it was time for me to get more rest. [. . .] I'm still theoretically employed but I have a feeling that my job will be gone by the time I return. [. . .] I've seen it in my own job: my boss never hires women with small children. He always says it's better if the kids are preschool age, like at least six or so. (Interview: Gdańsk 2007)

Rather than making 'irrational' decisions to delay motherhood and drive the aging Polish nation deeper into a demographic hole with confused priorities favoring careers over motherhood, women interviewed in this study consistently expressed a strong sense of responsibility toward their families and their children. As Renata, a 29-year-old environmental engineer poignantly observed, 'It's easy to birth ten kids but then they are reared in poverty. Everyone is responsible for their own kids and their own decisions of what they can manage and what is in their best interest, the family and their children'. Women consistently preferred more children if state policies supported their efforts to reconcile work and family. In this study's (predominantly middle class) sample, women wanted to 'self-realize', be treated fairly and decently, and feel materially stable to ensure their and their families' well-being. Women used the following expressions to convey this point: 'give women security that they can return to their jobs', 'protection of women at work', 'guarantees that women won't be fired', 'stop the practice of firing women from their jobs', 'legal protection of women in employment', and so on.

State's labor and health policies continue to offer no protection for women in the job market. The main 'Profamily Program' response from the Polish state came in 2005 when it introduced a one-time baby bonus known as *becikowe* – literally, money for a baby blanket – of 1000 złotych (approx. $337) per newborn. *Becikowe* was restricted in 2009 to only women who began pregnancy monitoring by a gynecologist not later than the tenth week, thereby drastically narrowing the eligibility for the benefit.[7] Women in this study deemed the payment 'absurd' and 'laughable' given the magnitude of financial and employment difficulties they face. Since the baby bonus failed to stimulate births, the Polish Minister of Finance is debating eliminating it altogether in 2011.[8]

In the end, women were highly aware of the 'demographic crisis' and many worried about the future of their retirement pensions and the aging of the society, but almost none felt compelled by the state's calls for increased fertility because of the ubiquitous perception that the state is not doing anything to facilitate motherhood and employment, and the widespread awareness that the government had dismantled the many social service programs that had been in place under state socialism.

Between a rock and a hard place

To balance personal desires, cultural expectations, and economic constraints, women in Poland walk a fine line between the expectation and a desire to have one child, and an increasing difficulty to have another – a bind that is underpinning the persistently low birthrate. The powerful cultural and Catholic stigma against voluntary childlessness makes childbearing less a free choice than a highly constrained one; still, Polish demographers are observing a slow increase in voluntary childlessness, although no data are available on the extent of this phenomenon (Fratczak 2004). But what is most striking in Poland, given the demographic crisis, is the existence of an equally powerful stigma against 'multi-child families', which are commonly referred to in the Polish vernacular as *patologia socjalna* – literally a 'social pathology'. The expression is used rather liberally and unselfconsciously in the political discourse, the media, and the professional world of social work (where much of the early research took place). It commonly conjures up 'reckless' reproduction in the context of rural and old-fashioned lifestyle, poverty, and perhaps alcoholism and domestic violence. How many children are deemed too many and therefore pathological? This can vary depending on the context.

The lower-income Polish and immigrant women, especially the Roma,[9] are typically the target of the pathology label, but the phenomenon is not unique to Poland. According to anthropologist Jane Schneider and sociologist Peter Schneider (1996, 8–13) who studied population decline in Europe, the politics of reproductive stigma originated in the late nineteenth century when limiting birth rate began to be associated with Western normative middle classness, thus separating so-called respectable and disreputable families based on their childbearing patterns. Soon, social pressures emerged to avoid the reproductive stigma of having 'too many' children. The participant observation work at the welfare offices in Krakow revealed that multi-child low-income families were often judged as exercising poor sexual control and showed lack of responsibility. Even in the eyes of some of the most dedicated social workers who tried to blame the Church for blocking access to sex education and therefore setting off 'reckless' reproduction, high fertility among the poor was still seen as draining limited welfare resources. Reproductive stigma vis-à-vis the poor and immigrants became normative when Europe began its demographic transition. As a result, the 'historical experience of limiting family size through sexual discipline – and in a context of social hierarchy – left behind a cultural residue that makes it easy to attribute any number of social ills such as backwardness, underdevelopment, and poverty to reproductive practices that elude consciousness or 'rational control'' (Schneider and Schneider 1996, 13).

But in the context of postsocialist Poland, the stigma of multi-childness appears to stretch across classes: while the 'pathology' label is liberally applied to the poor, the label of '*Matka Polka*' – literally, the Polish Mother – is used for middle class women. Both are stigmatizing and socially punishing in their implication that the woman has too many children; thus, a poor sense of judgment. The story of Krystyna, a social worker at a welfare office in Krakow, when she decided to have a second child is emblematic of this discourse. Krystyna and her husband, also a social worker, struggled financially on their state salaries but decided nonetheless that it was psychologically beneficial for their first child to have a sibling. Krystyna explained that when she was pregnant for the first time, her coworkers were approving and supportive, but when she got pregnant the second time she became the target of jokes. Her supervisor (herself married with one child) gave her the nickname 'Matka Polka' – a nationalist symbol of a self-sacrificing mother and a protector of the Polish nation derived from the Catholic cult of Mary. In the postsocialist era of the desirability of 'modern' lifestyles, *Matka Polka* is perceived in the popular discourse as an antiquated ideal of a religious woman. Krystyna recounted the story to me several years later in 2002 with a great deal of resentment, and explained that she tried not to get angry but nevertheless felt compelled to justify her decisions to her boss by saying that 'a child must have a sibling for proper socialization'. Krystyna's reasoning was promptly mocked by the supervisor who dismissed it as simply 'a typical way of thinking for a *Matka Polka*'.

It became clear in the course of this research that some women from middle and upper socioeconomic settings have experienced both types of pressures from their coworkers: to have a child, but not more than one child. Thus, having only one child appears to be not only a necessity for many women, but also a middle-class marker and a 'rationality' that separates the middle classes from the lower socioeconomic stratum, and alleviates some of the anxieties produced by the politico-economic instabilities that threaten the class standing of many. The association between modernizing postsocialist lifestyle and low fertility is increasingly more common in the popular discourse and in the media, despite the demographic alarm. In the case of Krystyna and her middle-class family, the stigma lay in Krystyna's embrace of an 'old-fashioned' religious female identity, namely that of a sacrificing Polish mother. She was the only one in a department of six women with more than a single child.

'The feminine genius'

In the context of Catholic nations, it is important to understand the religious pressures for unimpeded or 'natural' fertility, and the distinct religious rationality that underpins the policies and discourses around reproductive conduct. In response to declining fertility, the Vatican has been outspoken with warnings of irrational 'self-destructive tendencies' of individuals (read: women) who postpone marriage and childbearing for the benefit of careers. Other threats responsible for the 'demographic winter' are homosexuality and 'so-called new rights' for same sex couples, single parenthood, cohabitation, 'pro-divorce mentality', 'anti-birth mentality', and religious relativism, which makes people think that 'one religion is as good as another' (Lopez Trujillo 2003).[10] In an effort to revitalize reproduction as constitutive of womanhood, Pope John Paul II in his 1995 'Letter to Women' introduced the expression 'the feminine genius' arguing that women 'fulfill their deepest vocation' through motherhood, hence their reproductive capacity is their 'genius':

> It is thus my hope, dear sisters, that you will reflect carefully on what it means to speak of the '*genius of women*', not only in order to be able to see in this phrase a specific part of God's plan which needs to be accepted and appreciated, but also in order to let this genius be more fully expressed in the life of society as a whole. [. . .] *The Church sees in Mary the highest expression of the 'feminine genius'* and she finds in her a source of constant inspiration. Mary called herself the 'handmaid of the Lord' (*Lk* 1:38). Through obedience to the Word of God she accepted her lofty yet not easy vocation as wife and mother in the family of Nazareth. (John Paul II 1995)[11]

The Polish Church also condemned the 'evils of liberal pedagogy', and feminism as ideologies that promote individualism, thereby driving women to reject the self-sacrificing *Matka Polka* identity in favor of careers (Bartnik 2010). In particular however, the Church condemns contraceptive use and abortion (which at the level of the EU are defined as basic health services) warning against a 'contraceptive mentality'. The topic airs pervasively as the Church owns a number of television and radio channels, especially the ultra conservative *Radio Maryja*. The Church forbids all forms of contraception, including condoms and the withdrawal method, allowing only for periodic abstinence, popularly know in Poland as *kalendarzyk*, the 'little calendar'. Yet the demographic situation is making it obvious that Catholic women in nations such as Poland, Italy, Ireland, or Spain, manage and limit their fertility. How do Catholic women justify this practice?

Using contraceptives to manage fertility was of high priority for women in this study, despite the ban on abortion and having to resort to clandestine procedures, loss of subsidies for contraceptives, and the religious condemnation of contraception (Mishtal and Dannefer 2010).[12] They did not characterize their decisions as driven by selfishness or irrationality; instead they offered pragmatic rationales, in particular their sense of responsibility to their families over other considerations. The focus of women's narratives on 'rationality' and 'responsibility to their families' is symbolically highly significant, as women feel the need to defend their integrity as both mothers and rational actors in the face of religious and political discourses that denounce low fertility. However, in doing so, they are defining 'rationality' narrowly and potentially further stigmatizing women with more than a single child who then appear to lack this kind of 'rationality', as in how Krystyna was perceived.

Monika's (19-year-old university student, moderately religious) explanation of how she justifies contraception exemplified that way women reinterpreted what constitutes sin, and what falls outside of the Church's purview:

> Most of all I want to be able to control my own fertility, therefore I must protect myself against accidentally becoming pregnant. I'd never use the little calendar because it's simply not effective. The Church says that everything is a sin but the calendar, but I disagree. To me it's not a sin to use whatever method gives the needed protection. That's most important. [...] If my financial situation is suffering then that's all that matters and the Church has no impact in such situations; these are unrelated topics. (Interview: Gdańsk 2007)

In a direct address of fertility decline entitled, *Declaration on the Decrease of Fertility in the World*, the Vatican again argued for the protection of the heterosexual family against the dangers of the 'so-called new rights'. As is the case elsewhere in Europe where geographic or religious minorities are cast as politically threatening (Schneider and Schneider 1996), Vatican's pronatalist discourse is also about race and class and the dangers that immigrant populations, especially Muslim, represent to the 'religious balance' in Europe.[13] In Poland, the *Rzeczpospolita* daily's headline in January 2010 asked, 'Is Christian Europe threatened by Islamization?' – the article, accompanied by a photo of a woman wearing a niqab, related Cardinal of Prague, Miroslaw Vlk's statement warning of the imminent 'Islamization' of Europe due to the demographic crisis among Catholics (Szymaniak 2010).[14]

Conclusions

This paper highlights how Polish women, predominantly educated and middle class, cope with the changing political, economic, and social landscape after the fall communism by postponing motherhood and limiting fertility. Indeed, Poland has experienced persistent 'lowest low' fertility for over 15 years, and a negative population growth for the entire postsocialist period. The successive conservative administrations and the Catholic Church responded with nationalist and pronatalist rhetoric blaming women for 'selfish' reproductive decisions and calling for higher fertility to 'save' the nation. This rhetoric is also used to justify restrictions on reproductive rights. Yet, Polish women are refusing to embrace the role of the 'responsible biological citizens' by making childbearing choices that continue the low fertility.

Far from irrational rejection of motherhood, Polish women are highly pragmatic when delaying parenthood in order to navigate the new dilemmas marked by job insecurity, widespread gendered discrimination in employment, limited childcare, and cultural pressures for having children but not 'too many'. Against the background of major cutbacks in social services of the pre-1989 era, women in this study speak of new sources of economic instability and concern. Discriminatory employer practices are especially threatening to pregnant women and mothers – all illegal labor practices used by businesses to exclude undesirable employees. The state has not been able or willing to protect women against discrimination.

The class character of the fertility decline in Poland is highly significant as well. The emergence of a single child as the middle class norm based on 'responsibility to the family' reflects not only a necessity for many women but also a class marker that separates the middle class 'rationality' from the 'pathological' reproduction of the lower stratum. This emergent middle class identity in turn functions to lessen some of the postsocialist class anxieties produced by the instabilities of neoliberal market that increasingly endanger women's economic footing. However, the narrowly-defined 'rationality' of an only-child further stigmatizes women who lack the educational or employment recourses or otherwise make choices that do not adhere to this form of rationality, and are punished with 'pathology' or 'Matka Polka' labels.

The role of the Catholic Church cannot be underestimated in maintaining the current situation. When the administration of the Kaczyński twins in 2007 debated improvements in childcare services to help women stay at work, the Church shut down the idea as anti-family, arguing that the only pro-family approach is for women to be full-time mothers and give up employment (Jackowski 1997). Given that work–family reconciliation policies have increased birthrates elsewhere in Europe (France, for example), the Church's position clearly demonstrates that of utmost importance for the Church is to dictate the *way* women should reproduce, not to stimulate fertility. The Church is using the demographic decline as an opportunity to enforce a particular Catholic morality centered on the 'proper' reproductive and sexual conduct, and the revival of patriarchal family structure, without regard to the realities of women's desires, expectations, or the economic circumstances of their lives. Indeed, in the context of postsocialist democratization, the Church has been able to act as 'a superb instrument of power for itself' (Foucault 1991: 107), acting via legislative mechanisms to insist on a distinct form of religiously-sanctioned reproductive conduct and morality.

The ineffectual response from the Polish state to women's predicaments shows a fundamental lack of interest and political will to protect women's rights, and to make work and family reconciliation possible. Rather than implementing motherhood-friendly policies such as prosecuting gendered discrimination in employment or re-instating subsidized childcare, the state has instead taken up the religious rhetoric of primacy of motherhood and the nuclear family in their failed Profamily Program.

The influence of religious rationalities and the Catholic Church as a political institution in nation-building has long been recognized in population scholarship as a factor in reproductive politics and patterns. Although demographers favoring the tempo effect in declining fertility have relied on the notion of economic rationality, demographers have also critiqued their own focus on economics, noting other rationalities that shape different fertility contexts, in particular those that are influenced by religious or cultural pressures (Caldwell 1976, 355). The narratives of women in this study demonstrate that economic rationality theories are also limited by their partial understanding of the Polish case where the material and economic realities that circumscribe women's reproductive choices are in fact underwritten by *gendered* inequities. In this sense, Foucault's gender-blind conceptualization of biopower fails to account for the pronatalist techniques' and discourses' gender-specific target and effect (Soper 1993, 29–30). The biopolitical approach to understanding fertility decline also encounters another limitation in the context of Poland: it underestimates the complex responses of individuals to biopower (Lock and Kaufert 1998, 8–9). While the state socialist biopolitics aimed to enable women to obtain education and enter the labor market (succeeding on both accounts), the postsocialist biopolitics that restrict reproductive rights and ignore the need for policies to facilitate women's desire to balance work and family, are failing in motivating women to reproduce. At the center of the postsocialist case are pragmatic women – for whom 'demographic crisis' and religious discourses ring hollow – dealing with real dilemmas experienced in the new politico-economic climate, in particular gendered forms of inequality in employment and healthcare. These dilemmas are often eclipsed by the rhetoric produced by the state, the Church and the media that blames women for low births. The focus on women and their presumably irrational non-reproduction as the locus of the problem, therefore obscures the gendered inequalities inherent in the new power relations of the neoliberal state and labor market.

This paper argues that the Catholic Church rationality and the historical context of regime change in 1989 are decisive in the making and the persistence of the Polish demographic crisis. Thus, arguments here propose a departure from broad economic theories and call for a shift to a closer gender, class, and institutional power analysis in fertility research. Such analysis is critical in the context of Poland where the renewed power of the Polish Catholic Church has had major consequences for reproductive health policies and the religious agenda in the Profamily Program and politics. Even though religious calls for unimpeded fertility are falling on deaf ears, the religious discourses on the responsibility of women to reproduce, and by implication, the irresponsibility of non-reproduction, are ubiquitous and take attention away from the true dilemmas facing Polish women. The case of Poland highlights how conflict between religious agendas and women's rights threatens social and gender justice.

Acknowledgements

The postdoctoral research was supported by the Charlotte Ellertson Postdoctoral Fellowship and the Heilbrunn Department of Population and Family Health, Mailman School of Public Health, Columbia University. The dissertation research was funded by the Fulbright Scholarship. Institutional Review Board approvals were granted by the University of Colorado and Columbia University. The author thanks the panelists and participants in American Anthropological Association session at the 2008 conference where this paper was originally presented, as well as the two anonymous reviewers for their helpful comments. The author is also grateful to the many women who agreed to participate in this study as well as the clinics' directors, healthcare providers, and support staff for enabling this project.

Conflict of interest: none.

Notes

1. *Statistical Yearbook of Poland*, 1989–2007. Warsaw: Central Statistical Agency, p.124.
2. UN Population Division. Total fertility rate. http://data.un.org/Data.aspx?d=PopDiv&f=variableID%3A54#PopDiv

3. The Magistrate for Family Affairs: Program: Profamily Politics of the State. Warsaw, 1999, p. 3.
4. It is unsurprising that a focus on religious family values has the effect of curtailing fertility as scholars have shown that a religious family ethos is linked with greater gender inequity in the family, which in turn drives women's preferences for fewer children (McDonald 2000).
5. 'Explaining the increase in inequality during the transition'. See website: http://www.worldbank.org.pl/external/default/main?pagePK = 51187349&piPK = 51189435&theSitePK = 304795&menuPK = 64187510&searchMenuPK = 304822&theSitePK = 304795&entityID = 000009265_3980901093351&searchMenuPK = 304822&theSitePK = 304795 [Accessed on 12/29/2010]
6. 'Distribution of income or consumption'. See: http://siteresources.worldbank.org/DATASTATISTICS/Resources/table2_7.pdf
7. See: http://www.becikowe.com/
8. Discussions about the declining fertility in Europe are prominent at the level of the European Union; however, the EU has no mandate to require particular social service provisions or any other policies aimed at stimulating births. Thus, each member of the Union devises its own work–family reconciliation policy.
9. Roma women have actually been the target of sterilization campaigns elsewhere in Easter Europe, in particular in the Czech Republic.
10. 'The Church perseveres in her mission'. May 17, 1995. See: http://www.vatican.va/holy_father/john_paul_ii/audiences/alpha/data/aud19950517en.html
11. Emphasis in the original.
12. Modern contraceptive use increased in Poland significantly (56% in 2007) and the use of less reliable methods such as periodic abstinence or no method has declined dramatically. Still, in one study 36% of births were from unintended pregnancies. After the 1993 abortion ban, illegal abortion became widely available in private medical clinics for high fees (Mishtal 2010).
13. 'Declaration on the Decrease of Fertility in the World'. Pontifical Council for the Family. Vatican, Feb. 27, 1998. See: http://www.vatican.va/roman_curia/pontifical_councils/family/documents/rc_pc_family_doc_29041998_fecondita_en.html
14. The perceived threat of 'Islamization' and other anti-immigrant and nationalist sentiments that have intensified in this region after the Soviet collapse, constitute a major subject matter, and cannot be adequately addressed within the scope of this paper.

References

Bartnik, C. 2010. 'Zło pedagogiki liberalistycznej'. *Nasz Dziennik*. October 11. http://www.radiomaryja.pl/artykuly.php?id = 107898

Bongaarts, J. 2002. The end of the fertility transition in the developed world. *Population and Development Review* 28, no. 3: 419–43.

Caldwell, J. 1976. Toward a restatement of demographic transition theory. *Population and Development Review* 2, no. 3/4: 321–66.

Domanski, H. 2002. Is the East European 'underclass' feminized? *Communist and Post-Communist Studies* 35, no. 4: 383–94.

Erikson, Susan L. 2005. 'Now it is completely the other way around': Political economics of fertility in re-unified Germany. In *Barren states*, ed. Carrie B. Douglass, 49–72. New York: Berg Publishers.

Fodor, E., C. Glass, J. Kawachi, and L. Popescu. 2002. Family policies and gender in Hungary, Poland, and Romania. *Communist and Post-Communist Studies* 35, no. 4: 486–7.

Foucault, M. 1980. *The history of sexuality*, Vol. 1. New York: Vintage.

Foucault, M. 1991. On religion. In *Religion and culture by Michel Foucault*, ed. J. Carrette, 106–9. New York: Routledge.

Fratczak, E. 2004. Family and fertility in Poland – Changes during the transition period. Paper presented at the PIE International Workshop on Demographic Changes and Labor Markets in Transition Economies, February 20–21, in Tokyo, Japan.

Henzler, M. 2002. 'Więcej dziadków niż wnuków'. *Polityka*, December 7, nr 49 (2379) http://archiwum.polityka.pl/art/wiecej-dziadkow-niz-wnukow,376758.html

Jackowski, J.M. 2007. Bitwa o rodzinę. *Nasz Dziennik*. October 17. http://www.radiomaryja.pl/artykuly.php?id=91802

Johannisson, E., and L. Kovcs. 1996. *Assessment of research and service needs in reproductive health in Eastern Europe – Concerns and commitments*. New York: Parthenon Publishing Group.

John Paul II. June 29, 1995. Letter of Pope John Paul II to Women. Vatican. http://www.vatican.va/holy_father/john_paul_ii/letters/documents/hf_jp-ii_let_29061995_women_en.html

Lock, M., and P.A. Kaufert. 1998. *Pragmatic women and body politics*. Cambridge: Cambridge University Press.

Lopez Trujillo, A. 2003. The family and life in Europe. Bologna: Pontifical Council for the Family. http://www.vatican.va/roman_curia/pontifical_councils/family/documents/rc_pc_family_doc_20030614_family-europe-trujillo_en.html

Mazur, Peter. 1981. Contraception and abortion in Poland. *Family Planning Perspectives* 13, no. 4: 195–8.

McDonald, P. 2000. Gender equity in theories of fertility transition. *Population and Development Review* 26, no. 3: 427–39.

McDonald, P. 2006. Low fertility and the state: The efficacy of policy. *Population and Development Review* 32, no. 3: 485–510.

Mishtal, J. 2009a. 'Matters of 'Conscience': The politics of reproductive health and rights in Poland. *Medical Anthropology Quarterly* 23: 161–83.

Mishtal, J. 2009b. Understanding low fertility in Poland: Demographic consequences of postsocialist neoliberal restructuring. *Demographic Research* 21, no. 20: 599–626.

Mishtal, J. 2010. The challenges of reproductive healthcare: Neoliberal reforms and privatisation in Poland. *Reproductive Health Matters* 18, no. 36: 56–66.

Mishtal, J., and R. Dannefer. 2010. Reconciling religious identity and reproductive practices: The church and contraception in Poland. *European Journal of Contraception and Reproductive Health* 15: 232–42.

Nowakowska, U., and M. Korzeniowska. 2000. Women's reproductive rights. In *Polish women in the 90s*, ed. U. Nowakowska. Warsaw: Women's Rights Center.

Pawlicki, J. 2005. 'Europa: w poszukiwaniu utraconego sensu'. *Gazeta Wyborcza*, June 23. http://wyborcza.pl/1,75515,2783195.html

Rabinow, P., and N. Rose. 2006. Biopower today. *BioSocieties* 1: 197–217.

Reher, D. 2007. Towards long-term population decline: A discussion of relevant issues. *European Journal of Population* 23: 189–207.

Rivkin-Fish, M. 2003. Anthropology, demography, and the search for a critical analysis of fertility: Insights from Russia. *American Anthropologist* 105, no. 2: 289–301.

Rose, N., and C. Novas. 2005. Biological citizenship. In *Global assemblages: Technology, politics, and ethics as anthropological problems*, ed. A. Ong and S.J. Collier. Malden, MA: Blackwell.

Schneider, Jane C., and Peter T. Schneider. 1996. *Festival of the poor*. Tucson, AZ: University of Arizona Press.

Soper, K. 1993. Productive contradictions. In *Up against Foucault: Explorations of some tensions between Foucault and feminism*, ed. C. Ramazanoglu, 29–50. London: Routledge.

Szymaniak, M. 2010. Chrześcijańskiej Europie grozi islamizacja? *Rzeczpospolita*, October 1. http://www.rp.pl/artykul/417184.html

The Economist. 2004. Screwing the brand names. November 16.

Reproducing Italians: contested biopolitics in the age of 'replacement anxiety'

Milena Marchesi

Department of Anthropology, University of Massachusetts, Amherst, MA, USA

In national and international discourses, Italians are often represented as a greying population failing to reproduce itself. Italian women are targeted for their very low birth rates, while migrant women are scrutinized for their 'excessive' fertility and abortion rates. These demographic concerns over differential reproduction reflect 'replacement anxiety' about the below-replacement rates of Italians and the replacement of Italians by immigrants. Demographic anxieties coalesce with the intensifying of Catholic 'vitapolitics' manifesting in the paradox of pro-natalist interventions coexisting with the curtailment of fertility-enhancing reproductive technologies. The children of migrants emerge in some population discourses as a threat rather than a contribution to the reproduction of the nation. Drawing on multi-sited ethnographic research in Milan, this paper examines how reproduction in contemporary Italy has emerged as a contested social, political, and moral issue that invests Italian and migrant women in different ways, engendering different forms and terms of resistance and contestation. On what terms are subjects governed and called upon to govern themselves to be more 'rational' and 'responsible' reproducers of the nation? What subjectivities and local responses are engendered by the politics of reproduction in Italy? As different rationalities and notions of responsible reproduction circulate, ethnographic research sheds light on how anxieties over low birth rates are reappropriated and redeployed against the state, suggesting that subjects are not so easily governable by population and reproductive discourses. This research contributes to the literature on critical demography and the politics of reproduction and migration in the new Europe.

Introduction

The 'problem' of Italy's low fertility was the topic of a conference held in Milan in April 2007 and ominously titled 'Demographic Winter' (*Inverno demografico*).[1] The conference, co-sponsored by the City of Milan and the Italian Movement for Life, featured a prominent demographer of migration. According to the demographer, abortion was not the cause of Italy's low birth rates, but the 'abortion effect' was nonetheless significant as the 5 million 'missing children' due to abortion 'have subtracted lymph from a country that does not have much to begin with'. The demographer cautioned against interpreting the recent slight increase in Italy's total fertility rates (TFR)[2] as 'a true spring that renews itself by bearing fruits' because the increase in TFR from the 1990s' historic low of 1.2 to 1.4 a decade later is attributable to migration rather than to 'a natural effect'. Migrant reproduction, in the demographer's framing, although it may offer 'an injection of vitality', represents an artificial and even potentially dangerous demographic contribution: immigrants belong to 'problematic families,' the rate of population change is 'excessive' for the country to 'immunize itself,' risking the 'setting off of uncontrollable reactions'. Italians, meanwhile, 'play at the delay game', faced by structural difficulties, but also shaped by 'hedonistic tendencies,' such as the desire for 'that trip to the Caribbean'. In this discourse, an excessive migrant population and an Italian population whose lifeblood has been tapped into deficiency by abortion and low birth rates combine to threaten the nation's cultural and social reproduction.

The *Demographic Winter* conference is a local manifestation of a pervasive national and international discourse on the 'problem' of low fertility affecting a number of European countries. News coverage and cultural commentaries paint an almost apocalyptic scenario concerning the future of a greying continent no longer adequately reproducing itself. According to a recent editorial in the *Wall Street Journal*, for example, 'the greatest threat that Italians face is one of demographic self-immolation,' although 'Italy is not alone in committing demographic suicide' in Europe (Meotti 2010). Concerns with European fertility decline often combine its implications for the European social welfare model with anxieties over increasing populations of immigrants and their descendants (Castañeda 2008; Krause 2006; Sargent 2006).

In Italy, as in Poland, demographic anxieties are coinciding with intensifying 'vitapolitics' (Hanafin 2007) mobilized in defence of the embryo and converging in pro-natalist and pro-family politics and policies (see Mishtal, this issue). Legal scholar Patrick Hanafin defines 'vitapolitics' in Italy as a 'rigid top down' manifestation of biopolitics, 'a politics of entrapment in an imagined natural order,' a 'vitalist national narrative' that is threatened by women's reproductive choices (Hanafin 2007, 5). These vitapolitics are a manifestation of the resurgence of the Vatican's political power over the past decade, often deployed in defence of 'life'. The Vatican also echoes demographers and politicians in identifying cultural and moral dangers in Italy's and Europe's low birth rates[3] as it emphasizes the central importance to Europe's Christian identity of the 'traditional family' and of the sacredness of 'life'. These vitapolitics converge with secular and diffused forms of biopolitics that mobilize scientific knowledge in constituting embryos and foetuses as 'biopolitical subjects' and 'new forms of life' (Kaufman and Morgan 2005, 328, 317).

The governing of life through power 'situated and exercised at the level of life' (Foucault 1978: 137), which characterizes the modern liberal state and is enacted through 'rational' technologies of governance that seek to engender 'particular categories of citizen-subjects' (Ong 2003: 6), does not go uncontested. This paper traces discourses and policies of reproduction and the responses engendered on the ground among Italian and migrant women active in reproductive and migrant issues in Milan: on what terms are subjects governed and called upon to govern themselves to be more 'rational' and 'responsible' reproducers of the nation in a time of 'replacement anxiety' and intensifying 'vitapolitics'? What subjectivities and local responses are engendered by the politics of reproduction in Italy?

To answer these questions, this paper examines the intersections of two paradoxes of demographic discourse in contemporary Italy: First, a growing legitimation of the 'problem' of low birth rates has re-asserted the population as an urgent matter of national, social, and cultural reproduction in Italy. Yet, concomitant vitapolitical concerns with the status and protection of the embryo and with the nature of the family define the moral parameters of possible pro-natalist interventions within the confines of the 'traditional' family and the protection of the embryo. The moral discourse on the family and the embryo were central to the debate over the regulation of assisted reproductive technologies (ART) that played out in Italy in the early-to-mid 2000s leading to the restriction of access to these technologies, a paradoxical situation in a country saturated with demographic anxiety.[4] A second paradox concerns another facet of pro-natalist discourse in Italy: while there is a broad (although not universal) consensus that Italy needs more children, it is native Italians who are being hailed in demographic discourses. The reproduction of immigrants, in contrast, is construed as a problematic and even dangerous contribution to the nation. A common thread running through these discourses is the way the reproductive practices of women, both Italian and migrant, are marked as irrational, irresponsible, and even immoral. The articulation of these paradoxes, which in turn amplify and contradict each other, engenders resistance and contestation among Italian and migrant women, suggesting that reproductive practices and subjectivities are not so easily governable.[5]

The data on which the findings of this paper are based are drawn from nine months of multi-sited ethnographic research in the northern Italian city of Milan in 2006 and 2007. Policy documents and political discourse as well as ethnographic data were collected at national and regional conferences on immigration, demographic decline, and the family, as well as through participation in political activism centred on reproductive health and migrant rights, participant observation at an independent feminist family planning clinic, and the auditing of a three-month cultural mediation[6] training course. Data also include countless hours of informal conversations, over 250 hours of recorded meetings, discussions, and conferences, as well as semi-structured interviews with 21 Italian women and men engaged in reproductive politics and health and in migrant social services and activism.[7]

The 'problem' of the population

Political concerns over demographic statistics and attempts to intervene to alter them are not new in Italy. The birth rate of Italians has been a 'problem' for the State for at least a century, although the framing of the nature of the problem has varied. The fascist regime's support for the new science of demographic statistics, a discipline that made the population 'transparent' and amenable to being governed (Foucault 1978), paralleled its obsession with the regulation of the birth rates of Italians. The regime instituted pro-natalist policies, such as a tax on bachelors and the criminalization of contraception and abortion, deeming them 'crimes against the health and integrity of the stock' (Horn 1991, 585). Fascist pro-natalist policies, however, did not succeed in increasing the birth rate of Italians. Instead, the penetration of the fascist State into family-making practices engendered widespread resistance (see de Grazia 1992; Passerini 1987; Wilson 1996).

Following the demise of fascism, the strong association between the fascist regime and the discipline of demography led to a 'long "demographic silence"'[8] (Treves 2007, 47). In the 1950s and 1960s the 'problem' of the nation's population decline was replaced by anxieties over an excess of population, especially in southern Italy (Treves 2007, 50). Despite the fact that the understanding of the nature of 'the population problem' had reversed completely, there was significant continuity in the area of family and reproductive policy between fascism and the post-war Republican government (Caldwell 1989, 170). The criminalization of abortion and contraception, for example, remained on the books until the 1970s (Treves 2007, 52) when political movements, particularly feminism, succeeded in legalising divorce (1970), contraception (1971), and abortion (1978).[9]

The concern with overpopulation in Italy underwent another inversion in the 1990s with the re-legitimation of demographic discourses on the danger of fertility decline (Krause 2001). Demographers writing about Italy's low fertility rates blamed Italian women for their 'refusal to procreate'; one of Italy's most prominent demographers even compared Italian women to anorexic adolescents irrationally refusing food (Krause 2001). Until the early 2000s, politicians remained cautious of suggesting that any intervention that could be construed as pro-natalist and could thus trigger untoward historical comparisons. In the last decade, however, politicians across the political spectrum have embraced the problem of Italy's low fertility, often framing interventions in terms of family policy (Krause and Marchesi 2007).

In the mid-2000s, pro-natalist interventions became more blatant. The conservative coalition led by Silvio Berlusconi instituted a one-thousand euros 'baby-bonus,' a financial reward for women having a second child or beyond; however, only European citizens were eligible (Krause and Marchesi 2007). By 2007 it had become politically feasible for the centre-left government of Romano Prodi to establish the Ministry for Policies on the Family and to organize a three-day conference on the family, the first of its kind in Italy. The slogan for the conference: 'The family grows, Italy grows,' along with its highly unrepresentative logo of a family with three children,[10] point to the increasing acceptability of pro-natalist discourses and policies in Italy. At the conference, the Minister of Health, Livia Turco, celebrated this newly acquired legitimacy describing it as a 'cultural leap' for Italy (fieldnote: Milan, 2007).

Demographic rationalities and irrationalities

Theories of demographic change, most notably explanations for the demographic transition that began in the eighteenth century, have rested on a particular set of assumptions about modern subjectivity (Johnson-Hanks 2008; Greenhalgh 1996). The shift in reproductive practices that manifested in the transition from an equilibrium of high birth and death rates to one of low birth and death rates has been understood as the outcome of post-Enlightenment rationality and the inevitability of progress (Greenhalgh 1996, 27). However, the assumption that 'reproductive Westernization is good for everyone' (Greenhalgh 1996, 27) becomes complicated in the era of 'lowest-low' European fertility. As its long-standing status as a marker of modernity, progress, and rationality is qualified by charges that low fertility rates in Europe are in fact irrational and dangerous (Krause 2006), 'reproductive Westernization' is being cast in a new, and less favourable light. No longer an uncontested manifestation and symbol of rational progress, small families are increasingly seen as a threat to the survival of the West. 'Reproductive Westernization' remains the goal for non-Western societies and migrants and their descendants in Europe, while Western Europeans are increasingly targeted by pro-natalist discourses.

A country dying from low birth rates

Fertility rates in Italy have been classified in the 'lowest low fertility' range since 1992 and hit record levels in the mid-1990s when total fertility rates dipped below 1.2 children per woman (Caltabiano, Castiglioni, and Rosina 2009, 681), well below the 2.1 children per woman replacement levels for an industrialized population. Today, even with an increased TFR of 1.4 children per woman (Istat 2011), Italy continues to be categorized as 'a country with *persistent very low* fertility levels' (Caltabiano, Castiglioni, and Rosina 2009, 681). This low TFR reflects the fact that while 80% of Italian women over 40 have had at least one child, fewer have two children or more (Blangiardo 2010). Nonetheless, despite low fertility levels, the population of Italy continues to grow because of the contribution of immigration (Istat 2011) and increases in life expectancy.[11] Demographic data widely reported in the media underscore the fact that Italy's immigrant population is growing due to immigration and to the higher birth rates of migrants (2.31 children per woman) relative to that of native Italian women (1.32) (Istat 2010b). In 2009, 13.6% of births in Italy involved at least one immigrant parent (Istat 2010a, 2).[12] In 2010, immigrants represented 7.5% of Italy's population (Istat 2011).

Some politicians and demographers argue that the differential reproduction of migrants poses a threat to Italian society and culture. In a 2008 Parliamentary audition on the issue of low birth rates, for example, the Undersecretary of the Family, Carlo Giovinardi, warned that:

> This is a country that is dying from low birth rates, from the aging of the population, from a migratory flow so massive that it renders integration difficult since there is no longer an Italian society into which non-EU immigrants can integrate.... If this is the trend, in two or three generations, Italians will disappear.[13]

This 'demographic alarmism' (Krause 2006) over the racial and cultural reproduction of the nation represents the confluence of two kinds of 'replacement anxieties': anxiety about the below-replacement rates of Italian women and fears of the replacement of Italians by immigrants.[14]

Making 'responsible' reproducers

In response to 'replacement anxieties,' Italians, particularly Italian women, are directly and indirectly called upon to become more demographically responsible reproducers. Yet, what constitutes 'responsible' reproduction is far from a settled issue. Expert discourses of responsible reproduction assume the rational, planned, intentional reproductive behaviour expected of modern, post-demographic transition subjects. Other discourses circulate. Since the 1970s, Italian feminist thought has been critical of the equation of sexuality with heterosexuality and procreation and has embraced the term 'responsible reproduction' to refer to self-determination (see Bono and Kemp 1991). Politicians, particularly those on the right of the political spectrum, on the other hand, deploy the term in the context of disciplining reproductive behaviour, as they did in the debates over the legal restrictions of assisted reproduction approved in 2004. The Vatican's notion of 'responsible' reproduction refers to reproduction that follows Church law on matters of sexuality, the family, gender roles, and the regulation of births.

An example of expert interventions in the reproductive actions of Italians in the age of low birth rates is a project by the Italian Gynaecological and Obstetrical Society (Sigo) to educate women of reproductive age about their fertility. This project includes a YouTube video[15] in which the president of Sigo, gynaecologist Giorgio Vittori, discusses the importance of 'preserving one's fertility.' Vittori warns women not to prioritize careers too much: 'it is important to know that at 37 years old you may very well have a great job, a beautiful home, a great *trousseau*, but remember that the available oocytes are very, very few, remember it, think ahead.' The gynaecologist reminds women that 'it is important to maintain one's genital apparatus in perfect condition.'[16] In this formulation, responsible reproduction encompasses knowledge about contraception and sexually transmitted infections, but also about fertility decline over the life cycle. Italian women are thus called upon to engage in responsible sexual behaviour not for their general health and wellbeing, but with an emphasis on protecting their fertility.[17] Armed with this knowledge, women would presumably make more rational choices in balancing their career and family-making.[18]

For the Vatican, the only locus of responsible reproduction is the traditional family, and 'life' must be protected from the moment of conception.[19] The most explicit and influential elucidation of the Catholic position on 'responsible reproduction' is the 1968 papal encyclical *Humanae Vitae* in which Pope Paul VI addresses contemporary concerns with overpopulation. The pressing question for the Church at the time was whether 'the intention to have a less prolific *but more rationally planned* family' could justify the use of birth control. The *Encyclical* reaffirms the Church's stance by arguing that 'the exercise of responsible parenthood requires that husband and wife... recognize their own duties toward God, themselves, their families and human society', duties that preclude the acceptability of contraception.[20] The encyclical continues to be influential in Vatican discourses on reproduction (see Benedict XVI 2008). More recently, Catholic vitapolitics have manifested in the successful political involvement by the Vatican in the debates over assisted reproduction, particularly on the matter of the status of the embryo. The assertion of the embryo as a subject interweaves biology and theology.[21]

This brief mapping of some of the main articulations of what constitutes 'responsible' reproductive behaviour reflects the complexity of the politics of reproduction in Italy. Multiple discourses circulate, sometimes resonating with each other, at other times clashing. What constitutes responsible, rational, and moral reproduction in the time of demographic anxieties – and the question of responsible toward whom? God, the state, the nation, one's body, women? – remains up for grabs. The contradictions are fertile ground for critics of intensifying pro-natalist and pro-life biopolitics.

Taming the 'Far West' of assisted reproduction

Catholic doctrine and population politics at times articulate in paradoxical ways. In 2004, against the backdrop of 'replacement anxiety', the Italian Parliament approved legislation that severely restricts access to fertility treatment in Italy. Described as a law to regulate the 'Far West'[22] of assisted reproduction in Italy, the ART legislation introduced a moral discipline over who can access assisted reproduction, and which techniques are legally allowed: only 'stable' heterosexual couples can obtain fertility treatment in Italy while donor-assisted reproduction (using donor ova or sperm) was banned on the grounds that it threatens the harmony of the family and the identity of the future child (Marchesi 2007). In the name of protecting the 'weakest subject', the embryo,[23] from unscrupulous doctors and women made 'irrational' by their desire for motherhood, the ART legislation recognizes the embryo as an equal subject to the couple, bans cryogenic freezing, and limits the number of embryos produced by each in-vitro fertilisation (IVF) cycle to three, all of which must be transferred *in utero* regardless of their apparent viability.[24]

The Vatican was intensely involved in the debates over the legislation and in a subsequent popular referendum. Cardinals and other representatives of the Church, as well as Catholic politicians in favour of limiting access to ARTs, emphatically asserted the subjecthood of the embryo and accused women of being irrational and irresponsible in their desire to have a child 'at all costs'.[25] In these debates, women were addressed as active participants in practices that threatened the 'natural' family and social cohesion, but also as victims of their maternal desires and unscrupulous doctors. Politicians and activists opposed to restricting assisted reproduction, on the other hand, argued for the right to responsible reproduction as self-determination and for the right to access modern scientific technologies.

Negotiating biopolitics on the ground

The reproductive actions of women in Italy are targeted by contradictory reproductive policies and discourses. Marked as 'irrational' for not reproducing enough, Italian women are also accused of being irrational for pushing the 'natural' limits of reproduction by accessing fertility treatments. Feminist and reproductive rights activists, as well as other political actors, recognize and exploit these contradictions as they contest the emerging vitapolitics in Italy, while the reproductive actions of Italian women and men continue to escape demographic and moral incitements: a significant number (estimated at around 4000 per year) of Italian couples dealing with infertility travel to other European countries, particularly Switzerland and Spain, to access assisted reproductive treatments now illegal in Italy (Shenfield et al. 2010); and statistical data show that the birth rate of Italian women has not significantly increased in recent years. This suggests that intensifying biopolitical discourses have not, or not yet, succeeded in 'investing' or engendering the right kind of self-governing 'citizen-subjects' (Ong 2003, 6). Ethnographic research sheds light on how discourses and policies of reproduction are experienced and contested on the ground.

During the author's fieldwork in 2006 and 2007, the implications of the ART legislations were still reverberating among women on the left for whom the recognition of the subjecthood of the embryo in the context of ART regulation, an assertion widely understood to be laying the groundwork for making abortion illegal in Italy, was particularly offensive. Even as the ART legislation had not yet been approved, a centre-right Senator was quoted in the media arguing 'that having obtained the recognition of the principle that the embryo is a human being, we will now have to undertake a profound revision of the law on abortion in order to avoid a clamorous contradiction.'[26] Although such a revision has not been undertaken, indirect measures, particularly at the regional level, continue to take aim at the accessibility of abortion and at its moral significance. While public family planning clinics are increasingly de-funded, Catholic family planning clinics qualify for and obtain public funding.[27] The moral status of the embryo also emerges in policies that are seemingly unrelated to reproductive issues, such as the 2007 amendment to burial regulations in the region of Lombardy,[28] which mandated the burial of foetuses and embryos, regardless of age of gestation. Women undergoing abortions or suffering a miscarriage in Lombardy would be offered a 'choice': to privately bury the embryo or foetus, or to delegate burial to health authorities.

The author regularly participated in the meetings of a group founded in response to the approval of the ART legislation and focused on issues of reproductive health and rights. Participants discussed the intensification of the politics of reproduction and explored the possibilities and terms of an alternative politic. At a meeting following the approval of the 'foetal burial' amendment, participants reasserted their sense that one of the key fronts for feminist struggles was the ART legislation because of its recognition of embryo subjecthood. The times called for a deeper reflection because 'the moment the embryo becomes a subject, they can also impose a thought that is not your own. You impose your thinking because it would never occur to me. The imposition of a theological vision is dangerous, it's the sanctioning of a second-tier citizenship [for women]' (fieldnote: Milan, 2007). Daniela, an independent journalist, and one of the co-founders of the group, echoed her worry that 'Ratzinger [Pope Benedict XVI] constructs us from within.' Greta, a politician who had been active in women's politics for decades, argued that the foetal burial regulation could be traced back to the ART law: 'it continues to reproduce indirectly the notion that the conceived has juridical dignity' (fieldnote: Milan, 2007). Central to contesting these politics, then, was the refusal to recognize the embryo as a subject.

Viola, who participated in the group as a representative of a different feminist collective, decided with her group to organize a protest that would ridicule the very notion of the embryo's sacredness. The protest would be a mock funeral march for the unfertilized egg, to be symbolized by a coffin filled with bloody pads and tampons. The funeral march would wind its ways through the streets of downtown Milan on international woman's day. A flyer for the protest called on participants to gather at the central headquarters of the health administration offices (in charge of handling the burial paperwork) to turn in 'our used pads so that they be buried "with dignity"'. The irreverent protest was making fun of the rise of the sacralised embryo in Italian biopolitics. By giving a bloody sanitary napkin a funeral, the women of the feminist collective were casting the moral project of the embryo in an absurd light while reclaiming their bodies from both the state (by beginning the march at the headquarters of the health administration offices) and the Church, by appropriating religious symbols such as the staff, which was moulded into the shape of a uterus.

Contesting the state

The politics of the embryo justify the disciplining of both fertility-enhancing technologies and of abortion and contraceptive services. Caught at the intersections of moral and demographic discourses, Italian feminist women of different generations are critical of both the Vatican and the state. The book and documentary *Uno virgola due*, (One point two)[29] challenges the discourse that blames women for Italy's low fertility rate. According to Silvia Ferreri, the author and filmmaker, 'everyone is talking' about 'the problem of the empty cradles' raised by Italian President Azeglio Ciampi on International Woman's Day in 2004.[30] Ferreri summarizes this discourse as follows:

> the first who are considered responsible (not surprisingly) are women in their thirties. I feel implicated and for this reason the judgment weighs more heavily on me. Women in their thirties are in fact, for the most part, lazy women, concerned more with their hair than the future, they are *mammone*, they don't leave the home, are concerned with their careers (what career?), and not reliable enough to pull up their sleeves, take on the responsibilities that a family calls for, and raise some children. (Ferreri 2007, 22).

The book takes on this narrative of blame by examining the difficulties women who have children face in their workplace. The neoliberal retreat of the state from some of its welfare responsibilities and the spread of *precarietà* (precariousness) in employment contracts, whereby young workers are hired on short-term contracts without the labour safeguards enjoyed by previous generations (Molé 2010; Muehlebach 2007), provide plenty of fodder for this critique. Italian women's participation in the labour market is low by European standards and the gender wage gap remains significant (OECD 2008, 1–2).[31] Susanna Camusso, one of the participants at a roundtable discussion of the book held in Milan in 2007 cited the slogan '"*precarietà*" equals contraception' employed by some young Italian feminist activists as an example of the way the state fails to provide the right conditions for Italian women to have children (fieldnote: Milan, 2007). These critiques turn the tables back on the state, accusing it of irrational and irresponsible family and labour policies. Rather than challenge the alarmist discourse of low fertility, as many progressive Italians are wont to do, the book and the participants in the roundtable adopted its premises, that low fertility is a social problem, but critiqued the assumption that the problem lies with Italian women. Instead, they reframed low fertility as a further example of the failures of neoliberal and misogynist policies.

Disciplining migrant reproduction

While Italian women are targeted by pro-natalist discourses, migrant women's reproductive actions also elicit political, moral, and expert scrutiny.[32] Migrant reproduction represents a symbolic crossing of borders as 'the nation is reproduced through women's bodies' (Gal and Kligman 2000, 20). Because Italy functions under a *jus sanguinis*, or right of blood citizenship regime, the children of migrants born in Italy do not acquire Italian citizenship at birth.[33] Instead, migrants give births to foreign children. In this sense, migrant women 'subvert' borders (Yuval-Davis and Anthias 1989) by producing new immigrants without crossing national boundaries, threatening the body of the nation from within.

Migrant women are marked as a population of reproductive excesses, with higher fertility and abortion rates than their Italian-origin counterparts. Fertility rates among migrant women are approximately double those of Italian women.[34] The higher fertility associated with migrant women is celebrated by some as a resource for the country. Most often, however, the higher birth rates of migrant women are framed as a threat to the future of Italian culture, 'social cohesion,' and ultimately to the nation. In casual conversations people remark that 'immigrants are the only ones having children anymore.' Data from the national institute of statistics (Istat) on immigration rates and differential fertility rates regularly make headlines: 'Istat: "We have gone beyond 60 millions, record-breaking births from foreign parents. The children of immigrants fill 19% of the cradles of northern Italy;"'[35] 'Immigrants increase, they are 4 millions and 20% are undocumented.... A sharp rise is expected over the next few years, the numbers could triple with births;'[36] 'And now "the invasion" comes from the cradle. According to demographic projections, 10% of the children that will be born this year will be children of immigrants'.[37]

At the same time that anti-immigrant discourses warn of fertile foreigners overtaking Italians and their 'culture', migrant women are also the object of expert and moral discourses about their statistically higher abortion rate. Abortion rates associated with migrant women elicit moral concerns from the Church and calls for intervention from reproductive health activists and providers. These interventions are often based on the assumptions that the problem is a lack of education about contraception and of information about reproductive services available in Italy. A recent multi-regional government project is a case in point. Titled 'Prevention of voluntary pregnancy termination among foreign women,' the project is coordinated by the Ministry of Health, Ferruccio Fazio, who describes its aims as follows:

> The objective of the project is to reduce abortion rates among foreign women through better knowledge of the methods of responsible reproduction, of social and health services, such as family planning clinics, of laws in support of motherhood, of the diffusion of information on measures in support of motherhood and against the abandonment of infants. (ANSA 2010)

The notion that migrant women do not have knowledge of their bodies, pregnancy, and contraception is based on the assumption that migrant women have not been exposed to family planning education in their countries of origin. The globalization of family planning has made such assumptions problematic (see Ali 2002; Hartmann 1995). Moreover, Italians adopt medicalized contraceptive interventions at much lower rates than other Europeans, as the practice of *coitus interruptus* and the use of 'natural' family planning, supported by the Vatican, are widely used (Krause n.d). The education model ignores or is unable to address the structurally precarious location of migrants (see Calavita 2005), and particularly of migrant women in the labour market and in Italian society.[38] Migrant women are often employed as domestic workers. While they take care of Italian family members, young, old, and infirm, their own reproductive options are often severely limited (see Andall 2000).

Ironically, despite alarmist concerns about an 'alien' invasion of the cradles, the fertility rates of immigrant women have been declining.[39] In a recent paper, demographer Gian Carlo Blangiardo – the demographer at the 'Demographic Winter' conference – warns that migrant women's fertility is not the solution to Italy's demographic decline because the TFR of migrant women is also declining (from 2.5 in 2006, to 2.4 in 2007, to 2.05 in 2009) (Blangiardo 2010).[40] Thus, whether migrant women are represented as too fertile, not fertile enough, or as having 'too many abortions' (fieldnote: Milan, 2006), their reproductive actions are cast as a problematic demographic and moral contribution to the nation. Migrant reproduction is framed as excessive and thus in need of disciplining, or insufficient, and thus not worth the costs of immigration, and always suspect, a false spring that produces the wrong kind of fruits: more migrants.

Contesting demographic discourse

In conversations and interviews, a number of migrant women contested and reappropriated the discourses on irresponsible migrant reproduction. The experience of Dimah, a Moroccan woman who worked in a hospital as a nursing assistant and found herself pregnant from a short-term relationship, sheds light on the complexity of the articulations of demographic anxieties, notions of responsible reproduction, and of Catholic politics of life. When Dimah sought an abortion at a Milanese hospital she was directed to sit through a consultation with volunteers from the Italian Pro-life Movement, who are increasingly granted space in a public hospital wards where abortions are performed. Dimah described being aggressively counselled against having an abortion, an experience that upset her because she felt that, given her personal and employment situation, she had no choice but to terminate the pregnancy. Dimah recounted how she returned to her next appointment armed with her pay check of €900 a month and asked the volunteers: 'tell me how I can make it with a child'? Dimah concluded her narrative to the author by saying: 'it's something that can happen to anyone, even though I was careful. I hope that God will forgive me' (fieldnote: Milan, 2007). Dimah's comments reflect her desire to not be seen as irresponsible, both in how she became pregnant and in her decision to abort. By emphasizing her rational decision-making and the fact that she had been 'careful' and that even so pregnancy 'can happen to anyone,' Dimah countered stigmatized depictions of migrant women as 'irresponsible' and uneducated reproducers. Dimah's story also points to the way the reproductive actions of migrant women are subject to multiple levels of scrutiny and intervention, from the gaze of demographic statistics, to the expert concerns of medical and social service providers, to Catholic moral interventions.

Other immigrant women provided alternatives to dominant demographic and reproductive discourses. Bibiana, a Peruvian psychologist training to become a cultural mediator who had a young child with her Italian husband, referred to a study she had heard about on television which showed that Italian women's birth rate had significantly increased because of migrant women's domestic work. Bibiana used the study to recast the discourse on migrant women as a threat to the Italian nation into one that recognized their contributions to the reproduction of Italians. She had been a care worker in her early, undocumented years in Italy and described feeling imprisoned in the apartment where she took care of an Italian child:

> For me it was hard, the first year I cried a lot, a lot, Milena! Madonna! I was sad . . ., maybe Saturday or Sunday I . . . [was] with the child and I used to see young women like me who went out on their motorcycles, Italian girls, and I used to say 'darn!' What am I doing? Did I do the right thing coming here? [. . .] I didn't have any rest, I tell you that they had, some of the time they didn't give me the day off. Closed inside there, and some of the time they humiliate you, no? (Interview: Milan, 2007)

For Bibiana, migrant women in Italy sacrificed their freedom and their youth taking care of the children of Italian women, making it possible for Italians to have and raise children.

A similar appropriation and redeployment of the discourse on migrant reproduction was enacted by a Milanese migrant activist network. In the 2007 Milan May Day parade, the group explicitly challenged the discourse on the threat posed by the number of migrants in Italy and redeployed it in order to stake a claim to citizenship based on migrants' contribution to the reproduction of the nation. Speaking through a loudspeaker from an 'immigrant float,' Eva, one of the leaders of the group, argued: 'we are the ones who take care of your children, we are the ones who take care of your parents, we are the ones that are increasing the birth rate, if we weren't here, there would be no one to keep this country going' (fieldnote: Milan 2007).[41]

Migrant women's responses to the dominant discourse on reproduction aimed at them reveals the purchase of ideas of 'responsible' and 'irresponsible' reproduction and suggest that the biopolitical discourse on the threat and potential that migrants hold for the reproduction of the Italian nation can be re-deployed as the basis of claims for citizenship in Italy. Bibiana's argument that migrant domestic workers enable Italian women to reproduce, and the migrants' network's claim to citizenship on the basis of carework and higher birth rates, both mobilize discourses on migrant women's sacrifice for the Italian nation and redeploy them to stake 'counter-biopolitical' claims to citizenship (Comaroff 2007).

Conclusion

In the 1990s and 2000s, as low fertility among Italian women coincided with a rise in migration, governing the population gained new urgency. This paper has traced the discourses on reproduction and the population in Italy and the terms and forms of resistance they engender. Women in Italy are called upon to negotiate contradictory expert, political, and moral interventions aimed at shaping their reproductive conduct. These contradictions can manifest themselves in seemingly paradoxical political outcomes: despite intensifying discourses on the threat to the nation of low birth rates among Italians, the 'irrational' and even "immoral" reproduction of women who refuse to procreate shares the stage with debates on the need to regulate the 'irrational' desire for motherhood among women accessing fertility treatments. Migrant women are also targeted by contradictory discourses: they are marked reproductive subjects because of their higher fertility *and* abortion rates compared with Italian women, and even because of fertility decline over time.

Populations differentiated by nationality and reproductive statistics are called upon to act according to divergent reproductive rationalities and different understandings of what constitutes 'responsible' reproduction. The subjects of these discourses, both Italian and migrant in origin, have seized upon contradictory, if distinct, discourses and policies to contest intensifying biopolitical interventions in their lives. This finding supports anthropologist Rhoda Kanaaneh's reflection that 'population policies are frequently unsuccessful, at least by their stated goals . . . The desired production of "manageable" subjects often seems to remain elusive' (Kanaaneh 2002, 27). Both Italian and migrant women contest the moral framings that represent them as irresponsible, irrational, and even immoral reproducers, rejecting both the 'rigid top down' (Hanafin 2007, 5) vitapolitics of the Vatican and the self-governance promised by expert discourses. Understanding the effects of discourses on reproduction and the population on men and women in Italy today cannot be achieved in isolation from the political and cultural influence of the Church on matters of 'life,' reproduction, and the family, nor can it ignore expert discourses on responsibility and choice. It is at the intersections, and through the articulations, of these contradictory discourses that subjectivities are shaped, and that the terms and forms of resistance are formed.

Acknowledgements

The author is grateful to Silvia De Zordo, co-editor and co-organizer of the 2008 American Anthropological Association session on which this paper is based, for constructive feedback and discussions. The author also wishes to acknowledge the productive and thoughtful comments of the discussant for the session, Dr Elizabeth Krause, and the constructive feedback of two anonymous reviewers.

Funding for the dissertation research on which this paper is based was made possible by a Wenner-Gren Foundation for Anthropological Research Dissertation Fieldwork Grant (No: 7477, 2006–2007). This research received ethical approval by the University of Massachusetts Internal Review Board Committee.

Conflict of interest: none.

Notes

1. The following description and quotes are drawn from the author's fieldnotes and digital recordings, Milan, April 4, 2007.
2. TFR refers to 'total fertility rates,' the average number of children per woman, or more specifically: 'the number of children that a woman would bear if she survived through the ages of childbearing and gave birth at the age-specific rates of the time period' (Johnson-Hanks 2008, 302). Birth rates, on the other hand, refer to births per 1000 people. Italy's birth rate for 2010 was 9 per 1000 (Istat 2011).
3. See for example Cardinal Angelo Bagnasco's recent speech to the Italian Conference of Bishops in which he warned that 'Italy is moving toward a slow demographic suicide' (Vecchi 2010).
4. Italy's low birth rates are obviously not the result of individual Italians' infertility, nor would access to ARTs significantly alter Italy's birth rates. The restriction of assisted reproduction is relevant here because it speaks to the parameters of what and whose reproduction is considered responsible and favourable to the reproduction of Italian society and of the nation. An illuminating comparison is the liberal and subsidized access to ARTs provided by the state of Israel in response to its own demographic anxieties (see Kahn 2000).
5. The author is grateful to an anonymous reviewer for this important point.
6. Cultural mediators are predominantly women and almost exclusively migrants who have lived in Italy for some years and work in the social services, including the health care sector.
7. All interviews were conducted in confidentiality by the author and all names are pseudonyms.
8. All non-English language materials and literature cited in this paper have been translated by the author.

9. The legalization of divorce and abortion were confirmed by popular referenda in 1974 and 1981 respectively.
10. The average number of children per woman in Italy in 2010 is estimated at 1.40 (Istat 2010c, 2). Families with three or more children constitute 10.5% of total families with children in Italy (De Luca 2010).
11. The author is grateful to Elizabeth Krause for this point.
12. In 1995, children born of foreign parents represented 1.7% of the births (Istat 2009, 3). Births by at least one migrant parent increased 10.4% between 2009 and 2010 (Istat 2010a, 3).
13. Undersecretary of the Family, Parliamentary Testimony, July 2008 Commission XII Affari Sociali, Audizione del sottosegretario di Stato alla Presidenza del Consiglio, Carlo Giovanardi, sugli orientamenti programmatici del Governo in materia di famiglia e di droga, Seduta di giovedi' 3 luglio 2008, http://new.camera.it/_dati/lavori/stencomm/12/audiz2/2008/0703/s000r.htm, 8–9.
14. These discourses and anxieties over the internal and external threats posed by differential reproduction are not limited to Italy, of course; see Kanaaneh (2002) for an ethnographic analysis of anxieties over differential birth rates between Arabs and Jews in Israel; Chavez (2004) for research on the discourse of Latina fertility in the United States, and Heng and Devan (1992) on differentialist pro-natalist policies in Singapore.
15. Uploaded September 2008, http://www.youtube.com/watch?v=_Na6H7hlFKI
16. In an interview, Professor Vittori explains: 'we choose YouTube to speak to young women who often do not even consider the issue of a future maternity, on the contrary, they are often only concerned with avoiding unwanted pregnancies, without realizing how delicate is the equilibrium that regulates feminine biology.' http://italiasalute.leonardo.it/Copertina.asp?Articolo_ID=9361
17. While women are the most common target of political and expert interventions in matters of reproduction and fertility, Italian men also come under scrutiny, both in the media and in the comments of their female peers. The Italian daily *la Repubblica*, for example, reported a 2006 demographic research with the headline: 'Few children, it's men's fault "They are afraid of responsibility"' (Monestiroli 2006). The representation of Italian men as *mammoni*, or Momma's boys who fail to leave home and start their own families, generates headlines in other European countries and across the Atlantic (see for example, articles by the *Guardian*: 'Why Italy's mamma's boys can't cut the ties,' January 20, 2010, http://www.guardian.co.uk/lifeandstyle/2010/jan/20/italys-mamma-boys-cant-cut-ties, the BBC: 'Italians 'slow to leave the nest,' February 1, 2005, http://news.bbc.co.uk/2/hi/europe/4227675.stm)
18. These arguments ignore the concerns over precarious employment and shrinking social services.
19. The Catholic Church has a long history of disciplining reproduction and family-making in Italy (see Kertzer 1994).
20. *Encyclical letter Humanae Vitae*, Pope Paul VI, 1968, http://www.vatican.va/holy_father/paul_vi/encyclicals/documents/hf_p-vi_enc_25071968_humanae-vitae_en.html
21. Pope John Paul II, for example, established intellectual institutions such as the Papal Academy for Life (*Pontificia Academia per la Vita, PAV*), which has held conferences on matters related to the biological and ethical status of the embryo since 1994.
22. Until the approval of the ART legislation in 2004, medically assisted reproduction was governed by government circulars and the Italian medical association's code of ethic. The lack of a national legislation was seen as a problem across the political spectrum and feminist groups sought legislation to protect women from the unregulated market. In the debates over the disciplining of ARTs, the notion of Italy being the 'Far West' of reproduction was successfully appropriated by those favouring restrictions to protect the embryo and the family.
23. See for example the speech of Senator Corrado Danzi, a member of the Catholic Udc party, who argued: 'We consider this a just law, because at every step, in each of its articles, in each of its words, it foregrounds the defense of the weakest subject' (Senato della Repubblica, seduta pubblica 506, Resoconto sommario e stenografico, 11 Dicembre 2003, pg. 11, http://www.senato.it/documenti/repository/leggi_e_documenti/raccoltenormative/15%20-%20Procreazione/SENATO/AULA/st506.PDF).

24. Recent court decisions have found some of the more controversial aspects of the law unconstitutional: these include the limit on the number of embryos that can be produced and the injunction to transfer all embryos (ruling n.151/2009 of the Constitutional Court) (Legge 40, stop della Consulta 'No a limite di tre embrioni', *Repubblica.it*, April 1, 2009, http://www.repubblica.it/2009/03/sezioni/cronaca/viaggi-della-fecondazione/consulta-boccia/consulta-boccia.html); and the ban on donor assisted reproduction ('Legge 40, ancora un rinvio alla Corte Costituzionale,' *Corriere.it*, October 22, 2010, http://www.corriere.it/cronache/10_ottobre_22/fecondazione-costituzionale_f786e7fc-ddfb-11df-a41e-00144f02aabc.shtml).
25. Giuseppe Fioroni, Chamber of Deputies session 124, March 27, 2002, 22.
26. Senator Maurizio Ronconi, *Udc*. In *L'Unità* website, 10/12/03. http://www.unita.it/index.asp?SEZIONE_COD = HP&TOPIC_TIPO = &TOPIC_ID = 31220
27. Rapporto sull'attività dei consultori familiari accreditati 2003/2006 - Assessorato Famiglia e Solidarietà Sociale, Regione Lombardia.
28. Modifiche al Regolamento regionale del 9 novembre 2004, n. 6 'Regolamento in materia di attività funebri e cimiteriali', d'iniziativa della Giunta regionale.
29. A reference to Italy's TFR at the time of the book's publication.
30. 'Le culle vuote sono il primo problema italiano' ('Empty cradles are the foremost Italian problem'), Marzio Breda, *Corriere della Sera*, 3, March 8, 2004, http://archiviostorico.corriere.it/2004/marzo/08/culle_vuote_sono_primo_problema_co_9_040308027.shtml
31. The Organization for Economic Co-operation and Development (OECD) reports a 46% employment rates for women in Italy (OECD 2008, 1–2), which it attributes to the lack of publicly funded childcare and 'persistent labour market discrimination' (OECD 2008, 2). Additionally, the inequity that characterizes the labour market in Italy is matched by high gender inequity in house- and care-work (Cooke 2008; Ranaldi and Romano 2008, 22).
32. See for example the official summary of the 2010 report on abortion in Italy: Ministero della Salute. Salute delle donne. 'Presentata al Parlamento la relazione 2010 sull'applicazione della legge 194, che regola l'interruzione volontaria di gravidanza', http://www.salute.gov.it/saluteDonna/newsSaluteDonna.jsp?id=1159&menu=inevidenza&lingua=italiano
33. According to Italy's citizenship law (n.91/1992) children born to immigrant parents are eligible for citizenship only once they reach 18 years of age and only if they have been legal residents in Italy for the entirety of their lives. Considering the structural pervasiveness of periods of undocumented status for many migrants, many children of immigrants are in fact ineligible for citizenship once they reach age 18.
34. Ministero della Salute. 2010. Relazione del ministro della salute sulla attuazione della legge contenente norme per la tutela sociale della maternità e per l'interruzione volontaria di gravidanza (legge 194/78), pg., http://www.salute.gov.it/imgs/C_17_pubblicazioni_1312_allegato.pdf
35. 'Istat: "Siamo oltre 60 milioni," record di nascite da genitori stranieri. Superato il traguardo grazie agli stranieri. I figli degli immigrati riempiono il 19% delle culle del nord Italia,' la Repubblica.it, 23 June 2009.
36. 'Aumentano gli immigrati, sono 4 milioni e il 20 per cento è irregolare. Secondo i dati del rapporto Ismu, gli stranieri sono il sette per cento della popolazione italiana. Si prevede un'impennata nei prossimi anni, i numeri potrebbero triplicare con le nascite,' laRepubblica.it, March 6, 2007, http://www.repubblica.it/2007/03/sezioni/cronaca/immigrazione/immigrazione/immigrazione.html
37. 'E adesso 'l'invasione' arriva dalle culle. Secondo le proiezioni demografiche il 10% dei bambini che nasceranno quest'anno sarà figlio di immigrati,' laPadania online, January 3, 2007.
38. Ferruccio Fazio, Minister of Health in Berlusconi's conservative government coalition, argues that abortion rates in Italy are attributable more to 'socio-cultural factors, rather than to economic ones' (ANSA 2010).
39. The convergence of the fertility rates of migrants with those of residents of the country of destination has been documented by demographers and scholars (see Chavez 2004).

40. For a review of research testing various hypotheses on the demographic behaviours of immigrants compared with non-immigrants in the country of destination see Milewski (2010, 299–301).
41. The group also called for automatic citizenship for children of migrants born in Italy.

References

Ali, K.A. 2002. *Planning the family in Egypt: New bodies, new selves*. Austin, TX: University of Texas Press.

Andall, J. 2000. *Gender, migration and domestic service: The politics of black women in Italy*. Aldershot: Ashgate.

ANSA. 2010. Aborto: Fazio, attivato piano prevenzione per donne straniere. *ANSA (Agenzia Nazionale Stampa Associata) Notiziario generale in italiano*, August 9.

Benedict XVI. 2008. Messaggio del Santo Padre Benedetto XVI al Congresso Internazionale 'Humanae Vitae: Attualità e Profezia di un'enciclica,' Rome, October 3–4. http://www.vatican.va/holy_father/benedict_xvi/messages/pont-messages/2008/documents/hf_ben-xvi_mes_20081002_isi_it.html

Blangiardo, G. 2010. La situazione socio-demografica della Famiglia Italiana (ieri, oggi e domani). Conferenza Nazionale Della Famiglia: Storia e Futuro di tutti – Milan, November 8–10. http://www.conferenzafamiglia.it/media/6545/blangiardo.pdf

Bono, P., and S. Kemp. 1991. *Italian feminist thought: A reader*. Oxford: Basil Blackwell.

Calavita, K. 2005. *Immigrants at the margins: Law, race, and exclusion in southern Europe*. New York: Cambridge University Press.

Caldwell, L. 1989. Woman as the family: The foundation of a new Italy? In *Woman-nation-state*, ed. N. Yuval-Davis and F. Anthias, 168–83. New York: St. Martin's Press.

Caltabiano, M., M. Castiglioni, and A. Rosina. 2009. Lowest-low fertility: Signs of a recovery in Italy? *Demographic Researchart. 23* 21: 681–718, http://www.demographicresearch.org/Volumes/Vol21/23/21-23.pdf

Castañeda, H. 2008. Paternity for sale: Anxieties over 'demographic theft' and undocumented migrant reproduction in Germany. *Medical Anthropology Quarterly* 22, no. 4: 340–59.

Chavez, L.R. 2004. A glass half empty: Latina reproduction and public discourse. *Human Organization* 63, no. 2: 173–88.

Comaroff, J. 2007. Beyond bare life: AIDS, (bio)politics, and the neoliberal order. *Public Culture* 19: 1.

Cooke, L.P. 2008. Gender equity and fertility in Italy and Spain. *Journal of Social Policy* 38, no. 1: 123–40.

de Grazia, V. 1992. *How fascism ruled women: Italy 1922–1945*. Berkeley, CA: University of California Press.

De Luca, M.N. 2010. 'Crescere senza fratelli. Il sorpasso del figlio unico'. *Repubblica.it*, November 23, http://www.repubblica.it/cronaca/2010/11/23/news/crescere_senza_fratelli_il_sorpasso_del_figlio_unico-9398539/

Ferreri, S. 2007. *Uno virgola due: viaggio nel paese delle culle vuote*. Roma: Ediesse.

Foucault, M. 1978. *The history of sexuality: An introduction, volume I*. New York: Vintage Books.

Gal, S., and G. Kligman, eds. 2000. *The politics of gender after socialism: A comparative-history essay*. Princeton, NJ: Princeton University Press.

Greenhalgh, S. 1996. The social construction of population science: An intellectual, institutional, and political history of twentieth-century demography. *Comparative Studies in Society and History* 38, no. 1: 26–66.

Hanafin, P. 2007. *Conceiving life: Reproductive politics and the law in contemporary Italy*. Burlington, VT: Ashgate.

Hartmann, B. 1995. *Reproductive rights and wrongs: The global politics of population control.* Boston, MA: South End Press.

Heng, G., and J. Devan. 1992. State fatherhood: The politics of nationalism, sexuality, and race in Singapore. In *Nationalism and sexualities*, ed. A. Parker, M. Russo, D. Sommer and P. Yaeger, 343–64. New York: Routledge.

Horn, D.G. 1991. Constructing the sterile city: Pronatalism and social sciences in interwar Italy. *American Ethnologist* 18, no. 3: 581–601.

Istat. 2009. La popolazione straniera residente in Italia al 1° gennaio 2009. Popolazione. Statistiche in Breve. *Istituto nazionale di statistica*, 1–18, October 8. http://www.istat.it/salastampa/comunicati/non_calendario/20091008_00/testointegrale20091008.pdf

Istat. 2010a. La popolazione straniera residente in Italia al 1° gennaio 2010. Popolazione. Statistiche in Breve. *Istituto nazionale di statistica*, 1–18, October 12. http://www.istat.it/salastampa/comunicati/non_calendario/20101012_00/testointegrale20101012.pdf

Istat. 2010b. Natalità e fecondità della popolazione residente: caratteristiche e tendenze recenti Anno.

Istat. 2010c. Indicatori demografici. Anno 2010. January 24, http://www.istat.it/salastampa/comunicati/in_calendario/inddemo/20110124_00/testointegrale20110124.pdf

Istat. 2011. Demographic indicators. Year 2010. http://en.istat.it/salastampa/comunicati/in_calendario/inddemo/20110124_00/demographicindicators2010.pdf

Johnson-Hanks, J. 2008. Demographic transitions and modernity. *Annual Review of Anthropology* 37: 301–15.

Kahn, S.M. 2000. *Reproducing Jews: A cultural account of assisted conception in Israel.* Durham, NC: Duke University Press.

Kanaaneh, R.A. 2002. *Birthing the nation: Strategies of Palestinian women in Israel.* Berkeley, CA: University of California Press.

Kaufman, S.R., and L.M. Morgan. 2005. The anthropology of the beginnings and ends of life. *Annual Review of Anthropology* 34: 317–41.

Kertzer, D. 1994. *Sacrificed for honor: Italian infant abandonment and the politics of reproductive control.* Boston, MA: Beacon Press.

Krause, E.L. 2001. 'Empty cradles' and the quiet revolution: Demographic discourse and cultural struggles of gender, race, and class in Italy. *Cultural Anthropology* 16, no. 4: 576–611.

Krause, E.L. 2006. Dangerous demographies and the scientific manufacture of fear. The Corner House, Briefing paper, no. 36. http://www.thecornerhouse.org.uk/

Krause, E.L. Forthcoming. 'They just happened': The curious case of the unplanned baby and the 'end' of rationality. *Medical Anthropology Quarterly.*

Krause, E.L., and M. Marchesi. 2007. Fertility politics as 'social viagra': Reproducing boundaries, social cohesion and modernity in Italy. *American Anthropologist* 109, no. 2: 350–62.

Marchesi, M. 2007. From adulterous gametes to heterologous nation: Tracing the boundaries of reproduction in Italy. *Reconstruction: Studies in Contemporary Culture* 7, no. 1. http://reconstruction.eserver.org/071/marchesi.shtml

Meotti, G. 2010. Italia R.I.P.: By 2050, 60% percent of Italians will have no brothers, no sisters, no cousins, no aunts, no uncles. *The Wall Street Journal.* Opinion section, September 7. http://online.wsj.com/article/SB10001424052748704206804575467313288165570.html

Milewski, N. 2010. Immigrant fertility in West Germany: Is there a socialization effect in transitions to second and third births? *European Journal of Population* 26: 297–323, http://www.springerlink.com/content/1830u23634730063/fulltext.pdf

Molé, N. 2010. Precarious subjects: Anticipating neoliberalism in northern Italy's workplace. *American Anthropologist* 112, no. 1: 38–53.

Monestiroli, T. 2006. Pochi figli, è colpa degli uomini 'Hanno paura delle responsabilità.' *La Repubblica*, March 24. http://www.repubblica.it/2006/c/sezioni/cronaca/padrif/padrif/padrif.html

Muehlebach, A.K. 2007. The moral neoliberal: Welfare state and ethical citizenship in contemporary Italy. PhD diss., University of Chicago.

OECD. 2008. *OECD employment outlook 2008 – How does Italy compare?* http://www.oecd.org/dataoecd/42/38/40904438.pdf

Ong, A. 2003. *Buddha is hiding: Refugees, citizenship, the new America*. Berkeley, CA: University of California Press.

Passerini, L. 1987. *Fascism in popular memory: The cultural experience of the Turin Working Class*. Cambridge: Cambridge University Press.

Ranaldi, R., and M.C. Romano. 2008. Conciliare lavoro e famiglia: Una sfida quotidiana. *Argomenti*. n. 33, Rome: Istat. http://www.istat.it/dati/catalogo/20080904_00/arg_08_33_conciliare_lavoro_e_famiglia.pdf

Sargent, C.F. 2006. Lamenting the 'winter' of French fertility: Politics, power, and reproduction among Malian migrants in Paris. *Curare* 29: 71–80.

Shenfield, F., J. de Mouzon, G. Pennings, A.P. Ferraretti, A. Nyboe Andersen, G. de Wert, V. Goossens, and the ESHRE Taskforce on Cross Border Reproductive Care. 2010. Cross border reproductive care in six European countries. *Human Reproduction* 25, no. 6: 1361–8.

Treves, A. 2007. L'Italie, de la surnatalité aux 'berceaux vides': Réalités, representations et politiques démographiques (1945–2005). *Vingtième Siècle. Revue d'histoire* 96: 45–61.

Vecchi, G.G. 2010. Bagnasco e il calo delle nascite; 'Verso un lento suicidio demografico'. *Corriere della Sera*, 21, May 25. http://www.corriere.it/

Yuval-Davis, N. and F. Anthias, eds. 1989. *Woman-nation-state*. New York: St. Martin's Press.

Wilson, P.R. 1996. Flowers for the doctor: Pronatalism and abortion in Fascist Milan. *Modern Italy* 1, no. 2: 44–62.

Islamic logics, reproductive rationalities: family planning in northern Pakistan

Emma Varley

Department of Humanities & Social Sciences, Lahore University of Management Sciences, Lahore, Pakistan

This paper explores the use of Islamic doctrine and jurisprudence by family planning organizations in the Gilgit-Baltistan region of northern Pakistan. It examines how particular interpretations of Islam are promoted in order to encourage fertility reductions, and the ways Muslim clerics, women and their families react to this process. The paper first discusses how Pakistan's demographic crisis, as the world's sixth most populous nation, has been widely blamed on under-funding for reproductive health services and wavering political commitment to family planning. Critics have called for innovative policy and programming to counter 'excessive reproduction' by also addressing socio-cultural and religious barriers to contraceptive uptake. Drawing on two years of ethnographic research, the paper examines how family planning organizations in Gilgit-Baltistan respond to this shift by employing moderate interpretations of Islam that qualify contraceptive use as a 'rational' reproductive strategy and larger families as 'irrational'. However, the use of Islamic rhetoric to enhance women's health-seeking agency and enable fertility reductions is challenged by conservative Sunni *ulema* (clergy), who seek to reassert collective control over women's bodies and fertility by deploying Islamic doctrine that honors frequent childbearing. Sunnis' minority status and the losses incurred by regional Shia-Sunni conflicts have further strengthened clerics' pronatalist campaigns. The paper then analyses how Sunni women navigate the multiple reproductive rationalities espoused by 'Islamized' family planning and conservative *ulema*. Although Islamized family planning legitimizes contraceptive use and facilitates many women's stated desire for smaller families, it frequently positions women against the interests of family, community and conservative Islam.

Introduction

Under the leadership of President Ayub Khan (1958–1969), Pakistan was lauded as an example of the positive effects of population control initiatives (Lee et al. 1998, 953). However, following the widespread imposition of Islamization schemes, first by Zulfiqar Ali Bhutto and then, more critically, by Zia Ul-Haq in the 1970s and 1980s, Pakistan's family planning programs have experienced uneven success (Lee et al. 1998; Boonstra 2001, 5).[1] The pressures of regional political and economic insecurity, increasing religious conservatism and the rise of Islamist movements have led to diminished national and international funding for contraceptive services and provision (Lee et al. 1998, 957). This has been especially true for northern Pakistan, where Islamic conservatism is a key, albeit under-explored, factor in family planning implementation and uptake. Since the 1994 Cairo Conference on Population and Development (Gillespie 2004) and the 2005 Islamabad Declaration (IRIN 2005), Pakistan's population control lobby has made systematic efforts to formally involve and institutionalize the perspectives of leading Muslim clerics (IRIN 2005). The broadening of Pakistan's family planning services to include 'religion' follows from the productive ascription of Islamic tenets to family planning policy and programming in Egypt, Iran and Indonesia (Hasna 2003, 194; Tremayne 2004). Although it has not yet become a central component of family planning services in Pakistan, this approach represents a radical shift away from the secular tone of Pakistan's earlier and more successful family planning initiatives. It also reflects a transition in the demographic approaches which had once placed primary emphasis on the importance of 'conscious choice', the socio-economic advantages of family planning, and the availability of effective contraceptive methods for fertility declines (Underwood 2000, 111). Demographers have since come to embrace the idea that 'cognitive changes, and not economic considerations, are the primary driving force behind a decline in fertility' (Underwood 2000, 111). As part of organizational efforts to harness the 'social norms and shared values' (Underwood 2000, 111) underlying Muslim women's fertility regulation strategies, Pakistan's *ulema* have been called to an increasingly prominent role in the formulation and dissemination of family planning rhetoric.

Through ethnographic analysis of the introduction of 'Islamic' family planning in Gilgit-Baltistan, this paper analyses how family planning forms a point of engagement between multiple social forces that are themselves set amid fields of contestation. Over the past 30 years, Gilgiti women's bodies and reproductive capacity have been targeted simultaneously by secular development projects advocating 'smaller families' and intensive family- and mosque-centered religious pressure to bear more children. On one level, Islamist movements emphasize women's religious obligation to reproduce and enlarge the global *ummah* (community of believers). On another level, governmental and non-governmental Family Planning discourse situates women as bearing ultimate responsibility to curb population growth through contraceptive use or sterilization. In order to bridge this ideological divide and promote contraceptives as being religiously permissible, this paper examines how family planning organizations have used Islamic doctrine as a device through which a specific form of 'biopower' (Foucault 1978) operates to control and govern reproduction and sexuality. As an apparatus of biopower, family planning seeks to enforce specific fertility regulation practices and behavioral change at individual and population levels (see Rabinow and Rose 2006). This is accomplished through the articulation of religious criteria and conditional requirements that should ideally be fulfilled in order for fertility reductions to occur. 'Islamized' family planning is, therefore, a product of biopolitical 'strategies and contestations over problematizations of collective human vitality' and 'over the forms of knowledge, regimes of authority and practices of interventions that are desirable, legitimate and efficacious' (Rabinow and Rose 2006, 196). Moreover, by emphasizing the role of individuals in effecting population change, family planning cultivates 'modes of subjectification'

(Rabinow and Rose 2006, 203) that produce individuals who are expected to monitor, assess and regulate their reproductive strategies to meet 'rational' ideals. This paper explores how, by institutionalizing Islamic evaluative criteria and rationales as a means to encourage fertility regulation, family planning shifts from 'secular biopolitics' to what may be called 'Islamic biopolitics'.

Data concerning the Islamization of family planning is drawn from those portions of the author's longitudinal ethnographic research in Gilgit Town, the capital of Gilgit-Baltistan, which examined Sunni women's use of and attitudes towards contraceptives (2004–2005, 2010). Because the town's 70,000 residents are equally divided into Shia, Sunni and Ismaili factions, this location affords a unique opportunity to analyze how sectarian values intersect with the forces of health and development (see Varley 2010, 63–64). However, as the wife of a Gilgiti Sunni and with most fieldwork taking place during the 2005 'tension times' – ten months of low-intensity armed Shia-Sunni conflict, targeted killings, and protracted Army curfews (Varley 2010, 64–65) – research was predominantly restricted to Gilgit's Sunni community. Primary participants included approximately 30 urban Sunni women between puberty and menopause, who were recruited from across the socio-economic spectrum. However, social life in Gilgit Town is relatively egalitarian and predominantly comprised of a land-owning 'middle class'. Most participants had received basic education, were literate and belonged to extended virilocal family units. Several women earned modest incomes, for example, from traditional handicrafts, although none were financially independent. At one time or another, every participant had sought reproductive health or contraceptive services from the Directorate of Population Welfare or the Family Planning Association of Pakistan. Secondary participants were comprised of nearly 30 Ismaili, Shia and Sunni health service providers, family planning proponents and five members of the Sunni *ulema*. Participant-observation, semi-structured and open-ended interviews at household, community and institutional levels explored family planning policy and programming, Islamic doctrine and practices, and women's reproductive strategies and use of available health services. Secondary data included regional and institutional health surveys, policy manuals and Islamic texts concerned with issues of reproduction, fertility and contraception.

Because the non-governmental Family Planning Association of Pakistan provides the majority of secular and 'Islamized' family planning in Gilgit, the paper focuses on FPAP programs and publications that centralize Islam and apply religious criteria to authenticate and rationalize specific reproductive strategies. As ethnographies of Muslim societies attest, religious texts and resources exert considerable influence on matters related to everyday life (e.g., Lambek 1990; Ewing 1997). With the notable exception of Jeffrey, Jeffrey and Jeffrey's (2008) erudite and historically-informed analysis of Islamic revivalism, secular change and fertility in north India, further ethnographic research is required concerning the synergistic relationship between Islam, *ulema* and family planning proponents in South Asia. More broadly, Muslim women's use of Islamic doctrine to guide their reproductive and maternal health practices (e.g., Kanaaneh 2002; Sargent 2006; Tober, Taghdisi and Jalali 2006) also warrants additional attention. This paper adds to the literature by examining how Islamized family planning provides a cogent example of the contested ways that religious values and beliefs are written into reproductive regulation and strategies.

The Islamization of family planning

Compared with other parts of Pakistan, family planning in Gilgit-Baltistan is a relatively recent phenomenon. It has only been since the mid-1980s that Pakistan's Ministry of Population Welfare and the Family Planning Association of Pakistan (see FPAP 2004, 54; FPAP 2005, 2009) initiated broad-based contraceptive service coverage throughout rural and urban Gilgit-Baltistan.[2] In Gilgit Town, FPAP provides contraceptive services from its Family Health Hospital while the regional Directorate of Population Welfare distributes low- and no-cost contraceptives from government Rural Health Centers (RHC) and the District Headquarters Hospital (DHQ), the only tertiary-level referral hospital serving Gilgit-Baltistan. Subsidized contraceptives are also provided through FPAP and Population Welfare mobile service camps, and marketed and sold in Gilgit Town's private clinics, pharmaceutical dispensaries and retail shops by Greenstar, a national non-profit non-governmental organization.[3] As is true at the national-level (Karim 2011, 141), regional family planning services are managed primarily through the auspices of non-governmental organizations. At the time of research, the Family Planning Association of Pakistan provided the majority of in-town contraceptive services and 'Islamized' promotion as part of its reproductive and maternal health services, educational and economic community outreach initiatives.

Gilgit-Baltistan's low population density (12–20 per km versus 166 per km at the national level – see Government of Gilgit-Baltistan 2010, 4) means that high birth and fertility rates are not yet as onerous as those occurring elsewhere in Pakistan. The total fertility rate (TFR) is estimated to be 3.8 for urban areas (NIPS 2008, 37) such as Gilgit Town, while a regional survey in 1999 indicated an overall crude birth rate of 40.7/1000 (Rahman 1999, 11). However, increasing shortages in arable land in Gilgit-Baltistan's high-altitude communities constitute a growing challenge to population growth and are already correlated with rising rates of male-out migration. Such pressures, when combined with family planning promotion and outreach, have contributed to steadily rising contraceptive uptake rates. A 2008 health survey indicated that 48.1% of urban and 26.4% of rural women were currently using contraceptives (NIPS 2008, 48), although regional surveys fail to identify district-level variations or account for the reasons why Sunni, Ismaili and Shia communities demonstrate different fertility trajectories (Rahman 1999, 11; Varley 2010, 64).[4] For example, low contraceptive prevalence and high fertility rates among Gilgiti Sunnis, particularly in rural areas, stand in stark contrast to high contraceptive uptake and low fertility among religiously moderate rural and urban Ismaili communities (see Rahman 1999, 20–23; FPAP 2004, 3). Health personnel in Gilgit Town stated that while Sunni women make frequent use of reproductive health services, they make comparatively less observed use of family planning than Ismailis and Shias. In addition to the determinative influence of sectarian identity and beliefs, the quality of available health services and economic, geographic and cultural barriers to clinical access are implicated in reproductive service access and health outcomes (see Midhet, Becker and Berendes 1998; also Mumtaz and Salway 2005). Cultural and *ulema*-level resistance to family planning is also identified as a key obstacle to contraceptive coverage. As one physician noted, 'many women want less children, as do men, but clerics say family planning is a sin. They say, "How many children you have is God's concern, not yours"' (January 7, 2005).

While acknowledging that regional contraceptive prevalence has increased dramatically since the introduction of family planning services, Population Welfare and FPAP administrators stated that remaining 'health lags' and 'unmet need' among conservative Sunni as well as Shia communities warrant innovative and culturally appropriate solutions. For example, the inability of national Family Planning organizations to significantly decrease the national fertility rate using a 'two child' approach ('*doh bucchey acchey hain*'; 'two children are good') has led to an emphasis on the 'small family norm'. Given the seismic ideological, behavioral and socio-economic adjustments required for a 'two child' demographic transition, Gilgit's health proponents describe the 'small family' model as eminently more feasible and less controversial (see FPAP 2005, 7). In qualifying the 'small family norm' (see FPAP 2005, 6), an FPAP Community Organizer noted that contraceptive decision-making 'involves considerations of the family's economic status, health and access to health services' (May 11, 2005). Under the rubric of a 'small family' approach couples are therefore encouraged to determine family size according to their specific 'identified needs'. To this, the Population Welfare's Project Director spoke broadly of 'seeing the whole family holistically, including an understanding of children's well-being [and] education. It's about quality of life, not merely population statistics and numbers' (July 28, 2005). Such views are widely promoted through the medium of Urdu-language leaflets, community-based initiatives and provider-patient encounters.

When discussing institutional efforts to improve family planning's 'cultural acceptability', program administrators at the Directorate of Population Welfare and FPAP identified how establishing the Islamic permissibility of contraception has also been crucial to the success of regional contraceptive service provision and outreach. A primary element of this process includes the initiation of a series of workshops and seminars in Gilgit Town (FPAP 2004, 24), intended to operate as educational forums for Ismaili, Shia and Sunni clerics sympathetic to or interested in advocating family planning rhetoric, programs and outreach. In order to avoid provocation, workshop content is formulated so as to minimize the theological differences between each sect. Workshops use scriptural interpretation, doctrine and *fatwas* (jurists' opinions) common to each sect to establish the parameters of reproductive moderation. According to the FPAP's Community Organizer,

> We had to clarify the relationship between Islamic principles and family planning, so we met with *ulema* and *mohalla* [neighborhood] or village-level representatives. These included [Ismaili] *Mukkis*, [Sunni] *Maulanas*, and [Shia] *Ahol*, and we arranged training for them as well. We prepared workshops on family planning, about its benefits, what Islam says about family planning, spacing between children, mother's health [and the] economic challenges to growing families, breastfeeding and [also] sterilization. (May 10, 2005)

Evidence of the promulgation of family planning on Islamic terms is also provided by health promotion manuals that mimic the structure and content of Islamiyaat (religious studies) texts in order to offset or resolve the religious concerns of potential clientele. In recent years, four Urdu-language publications have been widely distributed by FPAP.[5] In each publication, moderate *ulema* extol the determinative influence of scripture (Qur'an), Prophetic tradition (*Hadith Al-Sunnat*), Islamic law (*Shariat*), Muftis' binding pronouncements (*hukm*) and jurists' opinions (*fatwas*) for assessments of the Islamic acceptability of contraception.[6] Between 1999 and 2003,

FPAP disseminated 8000 publications and promotional materials concerning Islamic positions on reproductive health (FPAP 2004, 24). Rather than provide materials directly to women or their families, FPAP employs a graduated dissemination strategy. Promotional resources are given first to leading clerics, politicians and community members and then to participating health care providers, community-based Lady Health Workers (LHWs), and volunteer male and female Health Guards throughout Gilgit Town. FPAP administrators explained that these individuals represented the most effective 'starting points' for the promotion of Islamized family planning at community- and clinic-levels. Participants confirmed that FPAP promotional materials have become accessible to and shared among women and their families through informal social channels. Because of the dearth of community health workers in Sunni neighborhoods and women's inability to attend mosque to hear clerics' sermons, these promotional materials represent an important opportunity for Sunni women to engage with and situate their reproductive strategies in FPAP's 'Islamized' terms.

Unlike FPAP's secular promotional materials, Islamized family planning uses the language of jurisprudence to promote contraception, a tactic that serves to segregate Islamic doctrine from the culture of Islam. Because of widespread consensus among Pakistan's *ulema* that permanent contraception such as surgical sterilization is prohibited by Islam (see FPAP 2004, 13), Islamized resources focus exclusively on establishing the permissibility of temporary methods. Although Gilgit's clinical services distribute only 'modern' allopathic contraceptives, such as the Pill, IUD, injectibles and Norplant (FPAP 2005, 10), FPAP publications focus almost exclusively on *'azl* (coitus interruptus), which is described as the only contraceptive explicitly permitted by the Prophet and recorded in the *Hadith Al-Sunnat* (Phulwari n.d., 11, 34–44; FPAP n.d. (a), 13).[7] For instance, in order to legitimate *'azl*, 'Islam and Family Planning' (Phulwari n.d.) first 'confirms' the method's permissibility according to *Hadith* and *fatwas*. It then uses analogous reasoning to argue if *'azl* is a 'lawful measure', other temporary methods may also be understood to be compatible with Islamic doctrine. And in ways that promote shared reproductive decision-making, *'azl* requires spousal agreement in order for it to meet Islamic requirements (Phulwari n.d., 62; FPAP n.d.(a), 5). Far more controversially, abortion is also approved prior to 'one hundred and twenty days gestation if the mother's health or breastfeeding ability is threatened' (Phulwari n.d., 85, 87).[8]

In ways that mirror Population Welfare and FPAP's emphasis on the 'quality of life' afforded by 'smaller families', Islamic family planning is defined to mean 'spacing out the birth of a child after the birth of a previous one, so that the abundance of children does not become a reason for distress for the parents' (Phulwari n.d., 9).

> ...family planning is not giving birth to a child after every two years. Rather, it is maintaining a reasonable time gap between the births of two children and planning childbirth according to increases in [families'] economic resources. This is the real meaning and purpose of family planning. (Phulwari n.d., 9–10)

This definition was frequently conveyed in interviews with FPAP staff and local physicians, who stated Islam had 'proven' that contraception was permitted if the ultimate intention was to increase the breaks between pregnancies (May 10, 2005).[9] FPAP publications advise that contraceptive use should be determined by means of processual, reflexive and religiously-informed decision-making. Contraceptive use is further predicated on the existence of individual- and collective-level factors that fit Islamic criteria. The reasons or 'logics' (Phulwari n.d., 39) used to justify contraception include poverty, economic instability and the lack of resources required to meet families' basic needs, preserving the health or breastfeeding potential of a mother, protecting very young or older women against potentially risky pregnancies, and avoiding illegitimate offspring from 'unlawful' relationships. Conversely, unethical or 'irrational' reasons may include women's fear for loss of beauty following childbirth (unless it incurs loss of affection by her husband), contraceptive use by the wealthy, or to avoid the births of daughters.

While secular promotion focuses predominantly on the socio-economic and health benefits afforded by 'smaller families', Islamized family planning materials also explore the moral and ethical precepts underlying contraceptive uptake and the decision-making process. 'Islam and Family Planning' (Phulwari n.d.) uses an interconnected chain of Qur'anic *ayats*, *Hadith* and analogous reasoning (*aqli o'naqli istadlal*), to explore the ethical principles that support or forbid contraceptive use (Phulwari n.d., 9). Jurists advise that decision-making should ultimately be guided by the maxim, 'if vice is the dominating factor, then the method will be illegitimate. If the predominating aspect is good, then the method will be considered justified' (Phulwari n.d., 12). Put otherwise, this enjoins that 'if there is a necessity, there is no harm' (Al-Musnad 1996, 163). FPAP therefore asserts that if Islamic requirements for contraceptive use have been met, family planning constitutes a 'lawful act' (*halal*). If contraceptive use fails to meet the broad range of criteria defined by moderate *ulema*, family planning is 'unlawful' (*haram*). By prioritizing Islamic dialogical and dialectical techniques, FPAP seeks to equip Gilgitis with the evaluative logics that will, in theory, factor in their appraisal of the ethical and situational pre-conditions to contraceptive use.

A second dialectical technique used to legitimize contraception involves identifying the scripture, *fatwas* and jurisprudence commonly used by conservative clerics to oppose family planning. These are then refuted according to a sequence of counter-*fatwas*, Islamic precedent and 'rational' reasoning.

> Another logic against family planning is presented [in] which the Holy Prophet (S.A.W.) said... 'Love a woman who loves and produces [children] a lot because, as compared to all *Ummats* [nations], I will be proud of the numeric strength of my nation'... The Prophet (S.A.W.) gave this order when economic equality existed alongside justice and Muslims only numbered a couple of hundred-thousand. They urgently needed manpower... Then make a decision yourself that today, when Muslims exist in their millions and homelessness, ignorance and unemployment is rampant, are there any guarantees of the resources required for life? (Phulwari n.d., 95–96)

In addition to literature addressing the Islamic permissibility of 'temporary' birth control, an FPAP-published Urdu-language brief entitled 'Precious Pearl' ('*Anmol Motee*', FPAP n.d.(b)) uses Qur'anic *ayats* to categorize the qualities associated with 'happy,' 'successful' and 'smaller' families. In essence, 'Precious Pearl' establishes the emotive and economic advantages of 'rational reproduction'. With its advice directed to a primarily male audience, fathers with too many children are described as less likely to provide their wives and children with good nutrition or a proper education, which in turn leaves families vulnerable to poverty and depression. The brief further infers that overburdened husbands will be unable to adhere to Islamic admonitions for wives to be given sufficient opportunity to breastfeed, or to have adequate breaks between pregnancies and thereby maintain their overall health (Phulwari n.d., 3).

In order to symbolically interweave contraceptive use with marital kindness and equality, 'Precious Pearl' highlights *ayats* that endorse men to treat women with respect (*ikhlaaq*).

Rather than promulgating new moral regimes, FPAP publications capitalize on considerable complementarity between Islamic, secular economic and biomedical rationales concerning contraceptives' suitability. Yet while the Islamic arguments foregrounded by FPAP publications appear to be informed by the secular cost-benefit ratios typically applied to gauge contraceptives' usefulness, they are almost wholly derived from formal jurisprudence and *Shariat* determinations. The format and content of FPAP publications also demonstrates remarkable similarity to widely-available Islamic 'social' and 'health' manuals. Such texts have served to promote the religio-political ideals of South Asian Islamic reform movements, such as the Jamaat-Ulema-Islam and Tablighi Jamaat (see Osella and Osella 2007). Reformist texts situate Islam in contemporary contexts, resolve Muslims' 'modern' concerns through the use of Islamic principles and precedent and, not unlike FPAP publications, distinguish between 'authentic' and 'inauthentic' religious, cultural and gendered practices. Given the strength and scope of Islamic social media and reformist religious publications across Pakistan, it is not surprising that the strategies successfully mobilized by reform movements have, to an important degree, served as a critical precedent for and been co-opted by FPAP.

However, in ways that may confound FPAP's policy objectives, the Islamic jurisprudence and *fatwas* cited in their publications rationalize the broadest spectrum of reproductive strategies; authorizing and endowing Islamic legitimacy to 'smaller' and 'larger' families, 'controlled' and 'uncontrolled' fertility. On one hand, reproductive moderation is valorized, while on the other, FPAP publications cite *Hadith* extolling reproduction's spiritual and economic merits; 'marry a woman who is very fertile and very loving' (Phulwari n.d., 44). The organization's fundamentally secular orientation has also proven problematic. Islamized family planning occurs alongside and is intended to complement FPAP's secular initiatives, such as the 'Girl Child Program' which incorporates skills training and home education (FPAP 2004, 25–28). To the dismay of conservative Sunnis, FPAP's in-town economic, educational and social outreach activities employ infiltrative 'social messaging' that advocates increases in women's social mobility, education, income-generation potential and decision-making autonomy. In the same way that women's bodies are marked as the grounds for change, matters of Islamic dress and gender segregation are also targeted. The FPAP's annual review of their 2004 'Empowerment of Adolescents' project notes:

> A positive impact was observed in the area where the adolescent girls' activities were implemented. The parents are now allowing their girls to go for education in the Project-based Home Schools and even permitted them to participate in different social activities. This positive change has been noted even in some areas which were considered more rigid toward social change and attitude for women and girls, thus the project has been able to shed the local custom of taking veil on [before] going out door[s] to participate in social activities. (FPAP 2004, 55)

Gilgit's family planning organizations mobilize two different yet similarly focused forms of reproductive rationality – one defined through formal doctrine, jurisprudence and interpretation, collective identity and shared health agencies; the other by human rights discourse and its attendant logics of modernity, progress and gender equity. Islamized family planning therefore remains fixed to a constellation of secular initiatives promoting women's agency, autonomy and equality in ways that are a politicized degree different from the gendered agencies bestowed by doctrinal Islam. Similarly, the Ministry of Population Welfare's organizational mandate stated that the objective of family planning programming is to 'bring about "social and economic development through rational choices about family size and reproductive behavior"' (Gul Khattak 2011, 150), confirming that secularism continues to supersede Islamized approaches.

The *Ulema*'s response to Islamized family planning

Gilgit's conservative *ulema* reject the broader secular logics associated with family planning and identify FPAP's Islamized approach as a false front for feminist social engineering. Clerics based at conservative mosques throughout Gilgit Town have charged that family planning leads to women's increased autonomy and the erosion of traditional family values. By extension, clerics have argued that contraception allows men and women to more easily indulge in 'illegal' extra-marital relationships (see Phulwari n.d., 19). In so doing, family planning is described as being interwoven with 'immoral' practices foreign to Islam and antagonistic to those traits and behaviors valorized by Sunnis as essential markers of religious and sectarian identity. As a part of their efforts to shift Sunnis toward 'authentic' Islam, clerics emphasize that frequent childbearing fulfills the requirements of faith and piety for both men and women. Such views are disseminated through public rallies, mosque sermons and conservative periodicals in which scholars respond to the arguments and 'rationalizations' employed by Islamized family planning.

For instance, where family planning proponents and moderate clerics state that temporary contraception is not explicitly disallowed by the Qur'an (Imran 1998, 1–2), conservative clerics counter-argue that nowhere in the Qur'an is contraception explicitly sanctioned. And when family planning proponents rely on *Hadith* authorizing *'azl* (withdrawal), conservative *ulema* respond that Muslims' responsibility (*zimwaree*) is to prioritize Qur'anic stricture ahead of secondary sources, including even the cherished *Hadith Al-Sunnat*. As part of their efforts to deflect attention away from the Qur'anic *ayats* and *Hadith* used to support contraceptive use, conservatists foreground the comparatively greater number of *ayats* and *Hadith* lauding childbearing and motherhood. Prominent among these is one *Hadith*, 'Marry the childbearing, loving woman for I shall outnumber the peoples by you on the Day of Resurrection' (Al-Albani 1979, 195 in Al-Musnad 1996, 164). This *Hadith* took on particularly dense sectarian meaning during the 2005 'tension times', when it was interpreted as speaking to Sunnis' perceived demographic vulnerability. Gilgit-Baltistan's most ardent critics of family planning were often the same clerics that extolled pronatalism as a tactical response to Shia aggressions.

The Tablighi Jamaat, a transnational Hanafi-Sunni missionization movement originating in Pakistan in the 1950s (see Metcalf 1998), was especially active in Gilgit Town's Sunni community during the 2005 'tension times'. Local and visiting Tablighi delegations, which also included women, spoke of the threats posed by Shia militancy for Sunnis, who hold minority status in Gilgit-Baltistan. [Shia sub-sects (Twelver, Ismaili, Nurbaksh) represent the majority of Gilgit-Baltistan's total population of 1.5 million, of which Sunnis comprise only an estimated 26% (Ali 1995 in FPAP 2004, 3).] Pronatalism has therefore featured as a 'rational' tactic by which Tablighi urge Sunnis to resist the 'spiritual degradation' posed by secular development and the demographic vulnerability incurred by sectarian conflict. As the young mother of two small children recounted, 'My mother-in-law has told me I have to produce more sons in order to protect the women [and] the family . . . from Shias' (April 28, 2005). Although barred from mosques, women were aware of the extremism inherent to pronatalist sermonizing.

> . . . mullahs say you should believe only in Islam and you can only stop having babies after twenty- to twenty-five children. The child you avoid conceiving could be a big mullah one day, they say, and you are also stopping all their future generations from being born. All Tablighi people are saying this, preaching this. (April 28, 2005)

Clerics' protests against family planning's influence on reproduction and childbearing facilitated health proponents' charges that conservative Sunnis were 'regressive' and 'backwards'. In trying to account for why Gilgiti Sunnis failed to use contraception in 'sufficiently' high numbers, local health care providers, many of whom are Ismaili and Shia, predicated Sunni women's apparent inability to easily access family planning on the 'harshly' uncaring and inequitable polemics of Sunni conservatism. However, by characterizing women in this way, providers tacitly upheld depictions of women as being recalcitrant, naive or lacking agency in their reproductive decision-making. Such perspectives coalesce with conservative Islamic discourse that frames women as intellectually deficient, lacking in 'reason' and 'logic', and too disempowered to make effective decisions. More problematically, health proponents also located women's use or non-use of family planning amid sectarian binaries. By first establishing equivalences between contraceptive use and religious moderation and, then by pointing to Ismailis' and then Shias' comparatively greater use of family planning, Population Welfare and the FPAP inferred – at least in the opinion of conservative Sunnis and *ulema* – that contraception use was analogous to the practices of 'non-believers' (*kufr*). At a time when religious identity and sectarian affiliation were paramount, such charges forced moderate and conservative Sunnis alike to take considerable care to disguise or divert attention away from contraceptive use.

Sunni women's reproductive strategies

Although Gilgit's health service providers, *ulema* and community leaders are the primary recipients of FPAP publications, participants confirmed they had direct access to or had been told about the contents of books such as 'About Family Planning: The Beliefs and *Fatwas* of the Scholars' (FPAP n.d.(a)) and 'Islam and Family Planning' (Phulwari n.d.). Women's health narratives demonstrated that the majority of participants had internalized the rhetoric conveyed by Islamized family planning. By citing Islamic precedent, such as specific *Hadith* and *ulema* rulings, women stated that with the consent of their husbands, they had the right (*haq*) to use contraceptives for the purposes of 'birth spacing', the length of which was to be determined by situational factors. Participants described how they also used FPAP publications in conjunction with Hanafi-Sunni 'social manuals' which aim to inculcate women with 'authentic' Islamic practices and affirm the gendered agencies bestowed by Islam (e.g., Madani 1999; Thanvi 2002). Participants used Prophetic (Islamic) medicine texts as well, many of which extol the health benefits afforded by birth spacing (Azimabadi 2004, 125–32). These findings suggest that Islamized family planning is paralleled and reinforced by South Asia's moderate Islamic reform movements, and confirm that traditional exegetical strategies frequently work to support and structure women's interpretation of the key Islamic messages conveyed by FPAP.[10]

Encouraged by the synthesis of regional family planning programming and Islamic doctrine, 'younger' participants (18–35 years) reported that they were more comfortable than 'older women' (35-menopause) to use contraception, or to ask their husbands to do so.[11] Their confidence was increased when local clerics emphasized moderate interpretations of Islamic doctrine to endorse temporary contraception. Rather than their husbands being defined as a primary obstacle to contraceptive use, women referred to their mothers-in-law's approval or disapproval when discussing contraception. More educated participants, especially those claiming some measure of autonomy because their husbands 'loved' them and heeded their wishes ahead of those of their mothers, claimed that their mothers-in-law were largely uninvolved with any overt discussion of contraception use. With Islamized family planning endorsing marital rather than collective decision-making, 'Ruqaiyah' discussed how her contraceptive use adhered to religious dictates because, 'such discussions are kept between [my husband] and me, and it doesn't involve my *Sas* [mother-in-law]' (August 16, 2005). Nonetheless, participants confirmed there was no easy way to separate issues related to fertility from the concerns and sometimes outright voyeurism of their in-laws. Although participants' mothers-in-law were generally supportive of women's use of Islam to expand their health-seeking agency, others actively intervened in the health of their daughters-in-law so as to uphold conservative reproductive trajectories. Religiously conservative mothers-in-law who self-identified as responsible for enforcing collective decision-making and 'traditional' reproductive rationalities, discussed their uncertainty concerning the types of 'awareness' their daughters-in-law might receive from health care providers. In order to exert authority over younger women's reproductive strategies, many insisted on being present during their daughters-in-law's clinical exams so as to be involved with or try to stop women's request for contraceptives.

Whether their use of family planning was decided privately or in partnership with husbands and family members, health and economic rationales for 'smaller families' dominated women's explanations for contraceptive use. Their considerations included the money required for children's education and family medical costs, the number of relatives available to help with childrearing, individual and collective beliefs concerning contraception's permissibility, and the degree to which women, their husbands and extended family subscribed to moderate or conservative interpretations of Islam. Family-specific reproductive traditions, which concern expectations of family size or the appropriateness of particular contraceptive practices, further influenced women's assessments of 'rational' decision-making and family size. Although participants acknowledged the importance of securing spousal consent, some women argued that because they alone faced the risks of pregnancy and childbirth, they were better suited to decide if their specific circumstances warranted contraceptive use. Given a regional maternal mortality rate ranging between 500 and 850 deaths per 100,000 live births (Mushtaq et al. 1999, 853; Rahman 1999, 12) and the very real dangers Gilgiti women face in their reproductive health, the majority of participants asserted that if pregnancy posed the threat of serious illness or even death, their husbands' and families' permission was not necessary. Other participants spoke against their families' and the conservative *ulema*'s apparent failure to appreciate the physical and mental challenges posed by childbearing. 'Samina', the wife of a conservative Tablighi Jamaat adherent, stated,

> The men and mullahs don't understand all the trouble, suffering and work women endure to be mothers. Being pregnant, giving birth, breastfeeding, caring for them, and all the other housework related to married life. It's none of their business if women [prevent] pregnancies or not, *awah nay* (right)? (May 4, 2005)

Despite voicing the desire for 'smaller families', among participants with only one or two children, family- and community-centered pressure to continue having children and sons in particular was often intense or unavoidable. In this respect, FPAP statistics are telling. Among 'Family Planning New Acceptors', 22% were the mothers of one or two children, 31% had three or four children, 28% had five or six children and 19% were already the mothers of seven or more (FPAP 2004, 18). Notably, none of FPAP's 'new acceptors' were childless, which provides compelling evidence of the pressure placed on newly-married women to avoid delays in childbearing. Observations of over three-hundred clinical visits at the FPAP's Family Health Hospital in 2005 and 2010 confirmed that the majority of patients, irrespective of sectarian affiliation, attended for reasons other than obtaining contraceptives, such as ante-natal or post-partum checkups, infertility or sexual dysfunction. Patients obtaining contraceptives were either the mothers of several children or sought family planning for 'health reasons'. In turn, health service providers at Gilgit's three hospitals confirmed that they usually only suggested the 'need' for family planning after women had delivered their third or fourth child. These findings indicate two things. First, that it is only with the agencies and entitlements achieved through motherhood and the births of sons that many women are able to act on their desire for contraceptives. Second, female providers adhered or deferred to the same logics concerning family size as many of their patients.

Research participants who were FPAP clients and the mothers of five children or less described how their use of contraceptives fulfilled organizational emphasis on reproductive moderation. In elaborating why, women pointed first to observed decreases in family size between the current and previous generation. Women then emphasized how their use of contraceptives hinged on economic and health-related factors and the permission and support of their husbands. Nonetheless, FPAP and Population Welfare representatives stated that Sunnis were still having 'too many' children. When these assertions were shared with them, Sunni women argued that while they were aware Islam permits contraceptive use under specific circumstances, FPAP publications fail to specify the exact size of a 'rational' family. As the following exchange between a middle-aged mother and her daughter attests, women's understandings of 'smaller families' were, in this context, relative.

> **'Yasmeen'.** I think five children is enough, but I've had eight. Changes are coming and nowadays people are not having more. Eight to ten was normal before, when there was no family planning.
>
> **'Zubya'.** Having three to four children is plenty! (June 2, 2005)

As was observed at all local reproductive health centers, FPAP health providers reaffirmed the value of 'smaller families' either through gentle joking, empathetic counseling, or sometimes outright exasperation at women's continued childbearing beyond 'reasonable' limits. For those patients for whom family size was decided by family members, physicians commiserated with women and expressed anger at the 'unreasonable' and 'unjust' health burdens imposed on women by husbands and in-laws.

While the majority of married participants were using contraception at the time they were interviewed, they spoke of the challenges posed by the wide array of stigmas and social anxieties related to family planning. For example, conservative clerics capitalized on women's fear of the 'sinful' and even 'genocidal' properties or moral and spiritual risks attendant on contraceptive use. For other women, the idea of contraception was less salient than the benefits provided by pregnancy and childbearing. This was particularly true for women who maintained close ties with village-based family, where reproductive rationalities were characterized by large families, cultural conservatism, religious fatalism, and politicized pronatalism. Pregnancy and childbearing afforded Sunni women and their families an undeniable abundance of material and relational benefits. Pregnancy's positive and 'rational' correlations with economic security, social status and the ability of childbearing to concretize marital and kinship bonds minimized the many presumed advantages of 'smaller families'. The symbolic and competitive advantage of fertility, the economic and political value of sons, and the enduring importance of family (*khandan*) served to uphold the value of pregnancy over non-pregnancy. This was especially true for participants who had been directly affected by or lost male relatives to sectarian violence. Clerics' emphasis on pronatalism as a replacement mechanism for those killed during the 'tensions' resonated strongly with these women and their families.

Against the backdrop of Shia-Sunni hostilities, women acknowledged that without their husbands' approval, their use of contraceptives or even abortion was 'unlawful' and risked being interpreted as an act of resistance not only against spouses and kin, but also the Sunni community itself. Under these circumstances, contraceptive use had to be inconspicuous or hidden. Cognizant of the insidious nature of neighborhood surveillance and gossip and in order to avoid catching the attention of local religious hardliners, or, exacerbating the animosities and power struggles inherent in many mother- and daughter-in-law relationships, women took considerable care to conceal the nature of their visits to family planning centers. If it was impossible to directly access family planning services, women made use of existing social networks to ask relatives or neighbors to procure contraception or abortifacients on their behalf. Participants described how the consequences for hiding contraceptive use included marital alienation, physical abuse or their being erroneously labeled as 'infertile' and thereby vulnerable to polygamous marriage or divorce. In turn, health service providers accepted that successful contraceptive uptake was predicated on patient privacy and confidentiality.

Conclusion

This paper has analyzed the uneven and densely political ways that the ideologies, practices and reproductive trajectories encompassed by Islamized family planning are conveyed and taken up. As part of the 'Islamization' of reproductive governance and regulation, Pakistan's population control lobby has institutionalized the notion that contraceptive use, at least in the minds of consumers, is predicated on Islamic precedent and contemporary *fatwas* authorizing family planning. At an institutional level, FPAP's objectives are shaped first by secular, women-centered and health-oriented mandates. But as part of recent efforts to promote the 'small family' norm and establish contraceptives' social acceptability, FPAP has used Islamic jurisprudence as a mechanism of health governance and reproductive regulation, or, 'Islamic biopolitics'. Islamic biopolitics can be defined as a type of population governance and health regulation that draws its impetus from and is invested in, particular religious logics. As a social modality, Islamic biopolitics encompasses an evaluative capacity that draws its strength from specific moral and ethical criteria, or, rationalizations which, in turn, are nuanced by sectarian and cultural vocabularies. Islamized family planning therefore represents the calculated use of religious rhetoric, analogous reasoning and jurisprudence to uphold and forward organizational policy and programming objectives. In effect, this serves to apply particular formulations of Islam to appraise, influence and govern women's reproductive and fertility management strategies.

As a manifestation of Islamic biopolitics, FPAP publications reflect the ethical, moral and rational sentiments associated with specific cultural contexts and interpretative traditions. In particular, Islamized family planning seeks to collectivize and socialize women's fertility regulation through the use of Islamic doctrinal precedent. In turn, by foregrounding the importance of dialogical, dialectical and analogical strategies for women's decision-making, Islamic practices, such as independent doctrinal interpretation (*itjihad*) and 'techniques of 'self-critique' (Huq 2008, 278), are written into FPAP's endorsements for women to measure their reproductive strategies against 'rational' standards and ideals. The role of Islamic evaluative criteria and value systems in women's decision-making processes aptly demonstrates the merging of biopolitics with 'ethnopolitics', defined by Rose (2001, 18) as 'the sentiments, moral nature or guiding beliefs... [that] provide the medium within which the self-government of the autonomous individual can be connected up with the imperatives of good government'. Rose argues that from an 'ethnopolitical' perspective, life (and in this case 'reproduction') becomes the 'object of adjudication' and the forum within which individuals judge and act upon themselves in accordance with, for example, notions of 'quality of life' (Rose 2001, 18).

As the paper also demonstrates, the Islamic logics used to uphold FPAP policy objectives draw their impetus from Islamic reform and missionization which, in similar ways, advocate for women's health and specific kinds of decision-making agency. Rather than representing novel moral regimes, Islamized family planning represents the redeployment of the reproductive rationalities debates espoused by South Asia's moderate Islamic reform movements. Participants' acknowledgement that they used both FPAP and reformist publications further suggests that the ameliorative effects women attributed to FPAP's Islamized approach may also be a result of larger processes aimed at clarifying the health 'rights' and agencies enjoyed by Muslim women and men. No less importantly, despite the health agencies and empowerment that arise from emphasis on women's right to contraception, 'Islamized' promotional materials fail to identify or propose solutions for the many 'irrational' obstacles preventing women from accessing family planning. Over-attention to the role played by doctrine in contraceptive use risks neglecting the reproductive strategies and decision-making dynamics that occur within the domain of culture and the popular practice of Islam. Such gaps are evidenced by FPAP's inability to provide culturally-appropriate strategies by which women can navigate the challenges posed by restricted social mobility, the pressure to deliver sons, husbands' refusal or non-support, family adherence to conservative Islam and the demographic anxieties incurred by Shia-Sunni hostilities.

This paper concludes that Islamized family planning is a fundamentally rationalizing project that builds its significance from religious, economic and biomedical evaluative logics. For Gilgiti Sunni women, contraceptive use is further nuanced by affective interplay between these logics, conflictive sectarian identity and family practices. Because Islamized family planning legitimates birth spacing and 'smaller families' not only through reference to Islam but also to issues of health or economic necessity, it provides women with considerable narrative and social leverage when justifying contraceptive use. Yet by endorsing sometimes sizeable shifts in decision-making and health practices, it has also placed women at-odds with the vested interests of conservative *ulema* and families oriented to 'traditional' reproductive strategies and rationalities. It is here, at the juncture of Islamized family planning, religious moderation and conservatism that the concept of 'irrational' reproduction emerges most forcefully. In one way, women described 'small families' as a 'rational' response to the risks posed by childbearing and the constraints associated with economic instability and extended family life. In another way, what family planning organizations label as 'irrational' reproduction instead reflected the wealth of economic, relational and spiritual benefits attributed to 'larger' families. This finding highlights the irony inherent in Islamized family planning that proposes to limit fertility in a region where 'excessive reproduction' has been exclusively defined through religion, and demonstrates that women's views of 'rational' or 'irrational' reproduction are not fixed. Notions of 'rational' and 'irrational' reproduction are instead generated by the interwoven ethical values and pragmatic expectations engendered simultaneously by Islamized family planning, Islamic reform movements, conservative sectarianism and family reproductive traditions.

Acknowledgements

The author wishes to thank Dr Michael Lambek and Dr Janice Graham for their support and guidance during the doctoral and post-doctoral fieldwork on which this paper is based. The author also wishes to acknowledge the feedback generously provided by Dr Milena Marchesi, Dr Silvia De Zordo and two anonymous reviewers. Research was granted ethics approval by the University of Toronto's Research Ethics Board (REB 12505; 2004–2005) and Dalhousie University's Office of Research Services (2010–2192; 2010). Research funding was provided by a SSHRC Doctoral Fellowship, an IDRC Doctoral Research Award (University of Toronto) and a Killam Postdoctoral Fellowship (Dalhousie University).

Conflict of Interest: none.

Notes

1. For the purposes of this paper, the term 'Islamized' family planning is derived from the broader concept of 'Islamization', which entails the reassertion of Islamic principles in Pakistan's socio-economic, political and legal spheres.
2. The Ministry of Population Welfare was devolved to Pakistan's provinces and territories in 2010 (Gul Khattak 2011).
3. Gilgit-Baltistan's population control lobby fails to incorporate the sometimes coercive regulatory and surveillance mechanisms characteristic of reproductive biopolitics in other areas of South Asia (see Van Hollen 2004). For example, regional family planning programs do not include the forced or surreptitious provision of contraceptives or surgical sterilization.
4. Because Ismaili and Shia communities have been comparatively more involved and invested in regional development than Sunnis, this has led to the clustering of benefits, awareness, behavioral and attitudinal change along sectarian axes. The Ismaili Aga Khan Health Service, Pakistan's (AKHS,P) achievements in reducing fertility and birth rates among Ismailis stands as a notable example (see Hertzman 2001).
5. These publications are 'Islam and Family Planning' ('*Islam aur Khandanee Mansoobah Bandee*'; Phulwari n.d.), 'About Family Planning: The Beliefs and *Fatwas* of the Scholars' ('*Khandanee Mansoobah Bandee sey Mutaliq: Ulmaie Deen key Afqar Fut'wah*', FPAP n.d.(a)), 'Precious Pearl' ('*Anmol Motee*'; FPAP n.d.(b)), and 'Family Planning in the Legacy of Islam' ('*Islami Miraz meh Khandanee Mansoobah Bandee*', Imran 1998).
6. In ways that may complicate Shias' and Ismailis' use of these resources, FPAP publications rely predominantly on the edicts generated by Sunni-Hanafi jurists affiliated with prominent seminaries (*madrasas*) and Islamic universities across Pakistan, India and also Egypt (FPAP n.d.(a), 7; Phulwari n.d., 81, 99–101). Family Planning in the Islamic Republic of Iran, on the other hand, uses Iranian and Iraqi-Shia jurisprudence (see Hasna 2003, Tremayne 2004, Tober, Taghdisi and Jalali 2006).
7. Other jurists note the Prophet's injunction for infants to be weaned only after two years and his prohibition of breastfeeding during pregnancy as evidence of his support of lactation as a 'natural' form of birth spacing (Phulwari n.d., 55).
8. Elective abortion is illegal according to Pakistani law.
9. Because of the divisive nature of inter-sectarian relations during and after the 2005 'tension times', providers deferred from speaking about Islam or issues of sectarian identity with their patients.
10. See Haniffa (2008), Huq (2008), and Marsden (2007) for ethnographic examination of the ways Islamic reformist texts are disseminated to and utilized by Muslim women in Bangladesh and Pakistan.
11. At the time of the author's research, six of 30 urban participants were pregnant. Four participants were actively trying to conceive, while the remaining 20 admitted to using some form of contraception.

References

Al-Musnad, Muhammad bin Abdul-Aziz. 1996. *Islamic fatawa regarding women*. Riyadh: Darussalam.

Azimabadi, Badr. 2004. *Prophetic way of treatment*. New Delhi: Adam Publishers and Distributors.

Boonstra, Heather. 2001. Islam, women and family planning: A primer. *The Guttmacher Report on Public Policy* (December): 4–7.

Ewing, Katherine Pratt. 1997. *Arguing sainthood: Modernity, psychoanalysis and Islam*. Durham, NC: Duke University Press.

Family Planning Association of Pakistan (FPAP) 2004. *Progress report: January 1999–September 2004*. Gilgit: Family Planning Association of Pakistan.

Family Planning Association of Pakistan (FPAP) 2005. *A brief report: Family planning on the roof of the world, FAP-Northern Areas*. Gilgit: Family Planning Association of Pakistan.

Family Planning Association of Pakistan (FPAP) 2009. *Annual report 2009*. Gilgit: Rahnuma-Family Planning Association of Pakistan.

Family Planning Association of Pakistan (FPAP) n.d.(a). *Khandanee Mansoobah Bandee Sey Mutaliq: Ulmaie Deen key Afqar Futo'wah* (*About family planning: The beliefs and fatwas of the ulema*). Lahore: Family Planning Association of Pakistan (Urdu language).

Family Planning Association of Pakistan (FPAP). n.d.(b) *Anmol Motee* (*Precious Pearl*). Lahore: Family Planning Association of Pakistan (Urdu language).

Foucault, Michel. 1978. *The history of sexuality: An introduction. Volume 1*. Trans. Robert Hurley. New York: Vintage.

Gillespie, Duff G. 2004. Whatever happened to family planning and, for that matter, reproductive health? *International Family Planning Perspectives* 30, no. 1: 34–8.

Government of Gilgit-Baltistan 2010. *Reform agenda-road to sustainable economic development*. Gilgit: Pakistan Development Forum.

Gul Khattak, Saba. 2011. The challenges of population policy and planning in Pakistan. In *Reaping the dividends: Overcoming Pakistan's demographic challenges*, ed. Michael Kugelman and Robert M. Hathaway, 145–59. Washington, DC: Woodrow Wilson International Center for Scholars.

Haniffa, Farzana. 2008. Piety as politics amongst Muslim women in contemporary Sri Lanka. *Modern Asian Studies* 42, nos. 2/3: 347–75.

Hasna, Fadia. 2003. Islam, social traditions and family planning. *Social Policy and Administration* 37, no. 2: 181–97.

Hertzman, Clyde. 2001. Health and human society. *American Scientist*, 89, November-December: 538–45.

Huq, Maimuna. 2008. Reading the Qur'an in Bangladesh: The politics of 'belief' among Islamist women. *Modern Asian Studies* 42, nos. 2/3: 457–88.

Imran, Abdul Rahmans 1998. *Islami Miraz Meh Khandanee Mansoobah Bandee (Family Planning in the Legacy of Islam)*. Lahore: Family Planning Association of Pakistan (Urdu language).

IRIN 2005. *Pakistan: Debating Islam and family planning*. IRIN Asia. http://irinnews.org/ReportID=28617; accessed April 22, 2010.

Jeffrey, P., R. Jeffrey, and C. Jeffrey. 2008. Disputing contraception: Muslim reform, secular change and family planning. *Modern Asian Studies* 42, nos. 2/3: 519–48.

Kanaaneh, Rhoda A. 2002. *Birthing the nation: Strategies of Palestinian women in Israel*. Berkeley, CA: University of California Press.

Karim, Mehtab S. 2011. Pakistan's demographic scenario, past and present: Population growth and policies, with lessons from Bangladesh and Iran. In *Reaping the dividends: Overcoming Pakistan's demographic challenges*, ed. Michael Kugelman and Robert M. Hathaway, 121–44. Washington, DC: Woodrow Wilson International Center for Scholars.

Lambek, Michael J. 1990. The practice of Islamic experts in a village on Mayotte. *Journal of Religion in Africa* 20, no. 1: 20–40.

Lee, K., L. Lush, G. Walt, and J. Cleland. 1998. Family planning policies and programmes in eight low-income countries: A comparative policy analysis. *Social Science and Medicine* 47, no. 7: 949–59.

Madani, Maulana Ashif Elahi. 1999. *A gift for Muslim women*. Lahore: Idara-e-Islamiat.

Marsden, Magnus. 2007. Women, politics and Islamism in Northern Pakistan. *Modern Asian Studies* 42, nos. 2/3: 405–29.

Metcalf, Barbara. 1998. Men and women in a contemporary Pietist Movement: The case of the Tablighi Jama'at. In *Appropriating gender: Women's activism and politicized religion in South Asia*, ed. Patricia Jeffery and Amrita Basu, 107–41. New York: Routledge.

Midhet, F., S. Becker, and H.W. Berendes. 1998. Contextual determinants of maternal mortality in rural Pakistan. *Social Science and Medicine* 46, no. 12: 1587–98.

Mushtaq, A., S. Shah, P. Luby, P. Drago-Johnson, and S. Wali. 1998. Survey of surgical emergencies in a rural population in the northern areas of Pakistan. *Tropical Medicine and International Health* 4, no. 12: 846–57.

Mumtaz, Z., and S. Salway. 2005. 'I never go anywhere': Extricating the link between women's mobility and uptake of reproductive health services in Pakistan. *Social Science and Medicine* 60: 1751–65.

National Institute of Population Studies (NIPS) 2008. *Gilgit-Baltistan demographic and health survey 2008*. Islamabad: National Institution of Population Studies, Government of Pakistan.

Osella, Filippo, and Caroline Osella. 2007. Introduction: Islamic reformism in South Asia. *Modern Asian Studies* 42, nos. 2/3: 247–57.

Phulwari, Mualana Alhaaj Shah Muhammad Jaffar. n.d. *Islam aur Khandanee Mansoobah Bandee: (Islam and Family Planning)*. Lahore: Family Planning Association of Pakistan (Urdu language).

Rabinow, Paul, and Nikolas Rose. 2006. Biopower today. *BioSocieties* 1: 195–217.

Rahman, Abdur. 1999. Northern health project (NHP): Northern areas baseline survey. Karachi: Abdur Rahman Associates.

Rose, Nikolas. 2001. The politics of life itself. *Theory, Culture & Society* 18: 1–30.

Sargent, Carolyn F. 2006. Reproductive strategies and Islamic discourse: Malian migrants negotiate everyday life in Paris, France. *Medical Anthropology Quarterly* 20, no. 1: 31–49.

Thanvi, Maulana Muhammad Ashraf Ali. 2002. *Bahishti Zewar (Heavenly ornaments)*. Karachi: Darul Ishaat.

Tober, Diane, M., M.-H. Taghdisi, and M. Jalali. 2006. 'Fewer children, better life' or 'as many as god wants?' Family planning among low-income Iranian and Afghan refugee families in Isfahan, Iran. *Medical Anthropology Quarterly* 20, no. 1: 50–71.

Tremayne, Soraya. 2004. 'And never the twain shall meet': Reproductive health policies in the Islamic Republic of Iran. In *Reproductive agency, medicine and the state: Cultural transformations in childbearing*, ed. Maya Unnithan-Kumar, 181–202. London: Berghahn Books.

Underwood, C. 2000. Islamic precepts and family planning: The perceptions of Jordanian religious leaders and their constituents. *International Family Planning Perspectives* 26, no. 3: 117–36.

Van Hollen, Cecilia. 2004. *Birth on the threshold: Childbirth and modernity in South Asia*. Berkeley, CA: University of California Press.

Varley, Emma. 2010. Targeted doctors, missing patients: Obstetric health services and sectarian conflict in Northern Pakistan. *Social Science and Medicine* 70: 61–70.

Programming the body, planning reproduction, governing life: the '(ir-) rationality' of family planning and the embodiment of social inequalities in Salvador da Bahia (Brazil)

Silvia De Zordo

Department of Anthropology, Goldsmiths College, University of London, UK

This paper examines family planning in Brazil as biopolitics and explores how the democratization of the State and of reproductive health services after two decades of military dictatorship (1964–1984) has influenced health professionals' and family planning users' discourses and practices. Do health professionals envisage family planning as a 'right' or do they conceive it, following the old neo-Malthusian rationale, as a 'moral duty' of poor people, whose 'irrational' reproduction jeopardizes the family's and the nation's well being? And how do their patients conceptualize and embody family planning? To answer these questions, this paper draws on 13 months of multi-sited ethnographic research undertaken between 2003 and 2005 in two public family planning services in Salvador da Bahia, where participant observation was undertaken and unstructured interviews were conducted with 11 health professionals and 70 family planning users, mostly low income black women. The paper examines how different bio-political rationalities operate in these services and argues that the old neo-Malthusian rationale and the current, dominant discourse on reproductive rights, gender equality and citizenship coexist. The coalescence of different biopolitical rationalities leads to the double stigmatization of family planning users as 'victims' of social and gender inequalities and as 'irrational' patients, 'irresponsible' mothers and 'bad' citizens if they do not embody the neo-Malthusian and biomedical rationales shaping medical practice. However, these women do not behave as 'docile bodies': they tactically use medical and non-medical contraceptives not only to be good mothers and citizens, but also to enhance themselves and to attain their own goals.

Family planning: A 'rational practice', a 'civic duty' or a 'right'?

> Family planning is a couple's right and the duty of all citizens living in a community. It is, above all, a means to turn human reproduction into a rational practice. It means allowing couples, whether they are rich or poor, to plan the growth of their family according to their desires [. . .]. The arrival of the first child can be postponed until the members of the couple have perfectly settled down, finished their studies and found a job. Contraception also allows controlling the fertility of the youngest and oldest

> women, whose children have more probabilities of suffering from foetal malformations. Family planning then improves the quality of the offspring, as it diminishes the number of physical and mental disabled people. (Coutinho 1998b, 81)[1]

Dr Coutinho, President of the ABEPF (Association of Brazilian Family Planning Organizations), Professor at the Department of Medicine of the Federal University of Bahia and founder of one of the first family planning clinics in Salvador da Bahia, explained with these words in his 1986 speech to the '*Escola de guerra*' (Military school) what family planning was and why it had to be supported. He concluded his speech by saying, 'The greatness of a country does not depend on the number of its soldiers. The only thing that counts is the quality of life of its inhabitants which depends on their physical and mental health and this can be found only in a planned family' (Coutinho 1998b, 81). Dr Coutinho delivered this speech before the military elite in a turbulent political period, when Brazil was transitioning from military dictatorship (1964–1984) to democracy and was struggling with the international debt crisis. In his view, the Northeast, one of the poorest and most highly populated regions of Brazil, where Bahia is located, had to be a priority target for family planning. In fact, fertility rates were still high there in contrast with the declining rates in the wealthy, industrialized Southeast,[2] where millions of poor Northeastern peasants were migrating. Moreover, there were high rates of infant mortality, and schistosomiasis and Chagas disease were endemic, leading to high rates of physical and mental disabilities among the population. In this context family planning was defended as an 'instrument of preventive medicine' (Coutinho 1998b, 82), the cheapest solution to the high infant and maternal mortality rates, as well as to poverty, overpopulation and urban criminality.

As was the case with most birth control supporters of his generation, Dr Coutinho conceived family planning as a weapon to fight against underdevelopment, a 'dispositif' (device), adopting a Foucauldian term (Foucault 1976), 'to turn human reproduction into a rational practice', in order to have healthy, productive citizens. Family planning emerges, in this eugenic perspective, as a 'biopolitics' (Foucault 1976) aimed at building a modern, developed nation by improving the quality of its population and creating good citizens. Physicians had the civic duty to educate poor people to 'rationalise' their desires and plan the size of their families according to their socio-economic and health conditions, and poor people had the civic duty to sacrifice or at least postpone their desire to have a family by adopting highly efficient contraceptives.

In the mid-1980s there were no public, state-funded family planning services. The production, advertising and selling of contraceptives had been legalized only in 1979, abortion was, and still is illegal except in cases of rape, maternal life risk and, since April 2012, also severe fetal brain injury, and tubal ligation was forbidden unless it was medically justifiable to prevent serious health problems. Despite its prohibition, tubal ligation was the most widely used contraceptive, followed by the Pill (BEMFAM 1986), which was freely distributed by the few existing private family planning organizations (such as the BEMFAM, Society for the Family's Wellbeing) funded by international organizations (such as IPPF, International Planned Parenthood Federation).[3]

This paper examines how family planning is envisaged in a completely different context by health professionals and patients of two public family planning services located in Salvador da Bahia, one of the major capitals of the Northeast, 20 years after Dr Coutinho made his speech. After the new Brazilian Constitution defined family planning as a right in 1988, the SUS (the free, universal health care system created after the restoration of democracy) started to provide free contraceptives and family planning education. In the 1990s, HIV-STDs prevention services were also implemented. As a result, Brazilian women today use more temporary contraceptives and less tubal ligation. At the same time, as a result of the implementation of the SUS and of the general improvement of life conditions, infant and maternal mortality rates have decreased considerably. However, they are still higher in the Northeast than in the Southeast, particularly among low income black people. In this context, do physicians, nurses and their patients still envisage family planning as a 'moral', 'rational' decision, the social/moral duty of all good parents/citizens, or do they conceive it as a 'right', a 'free choice'? And what is the experience of family planning users with the contraceptives provided by public services? Have they embodied the biomedical rationale that should supposedly orient them in their use?

This paper answers these questions using the perspective of a 'political anthropology of the body' and 'of health' (Fassin 1996, 2000), drawing on 13 months of multi-sited ethnographic research undertaken between 2003 and 2005 in two family planning centres funded by the SUS in Salvador da Bahia,[4] the third most populous Brazilian city, with a majority 'black'[5] population and high rates of unemployment. It focuses in particular on 'Gine-Salvador',[6] a private clinic and the only one providing every day, in 2003–2005, a free family planning service including family planning education.[7] Unlike most public health services, this clinic also had a surgical centre: therefore, it provided not only temporary contraceptives, but also tubal ligations and vasectomies, performed at only a few public hospitals. After briefly summarizing the history of family planning in Brazil, this paper examines the interaction between health professionals and patients and their respective views on family planning and contraception. It shows how discourses on the one hand on 'rights', particularly 'women's rights', and on the other hand on 'rationality', 'irrationality' and 'culture' shape medical practices and female patients' reproductive and contraceptive trajectories. Finally, it interrogates the extent to which it is possible to affirm that in the 21st century the bio-politics of the population of the past, aimed at limiting the fertility of the poor, have been substituted, as Rose (2007) argues, by the bio-politics of the 'self', aimed at enhancing individual choice and personal autonomy.

Family planning in Brazil: From eugenics to reproductive rights

When Dr Coutinho delivered his speech, fertility rates were declining very quickly (from six children per woman in the mid-1960s to four children per woman in the mid-80s) and female sterilization was the most used contraceptive in Brazil. Most tubal ligations were performed via caesarean sections in order to be medically justified,[8] with patients paying their physicians 'under the table' to carry them out. A Parliamentary Commission created to study this phenomenon in 1991 found that patients were not always given a consent form to sign before the procedure, and several poor and indigenous women had been sterilized against their will. Moreover, the Commission found out that in the poorest regions, such as the Northeast, local politicians provided free tubal ligations to poor women in exchange for a promise of a vote (Caetano and Potter 2004). Finally, some firms asked women to provide a sterilization certificate prior to commencing employment.

In light of this evidence, representatives of the black movement denounced family planning campaigns as eugenic programs aimed at limiting births among black people, who were and still are overrepresented among the poor. A survey carried out in the mid-1990s (BEMFAM 1996) showed that women's sterilization rates had increased (31% of married women in reproductive age using contraception were sterilized in 1986, 40% in 1996), while the Pill's use rates had declined (from 28.7% in 1986 to 20.7% in 1996). There were no statistically relevant colour/race differentials in the rates of sterilized women. However, women with a lower educational and economic level had undergone tubal ligation at a younger age compared with upper-middle class women, and did not use (because they had no easy access to) a wide range of temporary contraceptives (Berquò 1999). A heated political and scientific debate and the pressure on the government by the feminist and black movements[9] led to the promulgation of a Law on Family Planning in 1996 and to the improvement of free family planning services.

As a result of these efforts, Brazilians nowadays use many more temporary contraceptives, particularly the condom and hormonal injections, and less the withdrawal and the rhythm method, while the IUD is still used by a minority of Brazilian women (Ministério da Saúde 2008). During the last decade fertility rates fell below two children per woman, but female sterilization rates also decreased (29% of married women using contraceptives were sterilized in 2006), while the rate of vasectomies increased (from 2.6% in 1996 to 5.1% in 2006). At the same time, the discourse around 'sexual and reproductive rights' and 'gender equality' became dominant in political and media debates. However, safe abortion is still accessible only to upper-middle class women in private clinics, while unsafe abortion is a common phenomenon (Diniz and Medeiros 2010) and in Salvador it is the first isolated cause of maternal mortality, mainly among young, black women with a low educational level, living in peripheral areas (Menezes and Aquino 2009). Moreover, the rights discourse is used nowadays not only by movements supporting women's reproductive rights, but also by those supporting the 'rights of the foetus' (Morgan and Roberts 2009). Finally, recent studies have showed that low income women, who struggle with deep social and gender inequalities, and young, black women in particular, still face many difficulties in negotiating and using effective contraceptives (or in using them effectively) to prevent unintended pregnancies (Heilborn et al. 2006).

How do health professionals working in public family planning services envisage family planning, in this context?

Family planning in physicians' view: Structural problems and 'cultural' barriers

When asked to provide his definition of 'family planning', Dr Fábio,[10] a gynaecologist working at 'Gine-Salvador', responded:

> If we allow family planning, we prevent complications and constraints which can affect families. So you give people the right to choose to have as many children as they want, but not based on this view, that one could define as fascist, that the poor can't have any children, while rich people can have as many children as they desire [.] The poor [man] can say: I want to have five children. Very well . . . but he must know about the problems that may result from having more than one, two or three children. Sometimes his financial situation is not adequate but. . . there is also the emotional aspect: having one child is very important for many people, having a family is important for

> almost everyone. So they have the right to know how many children people can have and until what age they can have children. (Interview with Dr. Fabio, Salvador 2004)

Dr Fabio was from a far younger generation than that of Dr Coutinho and did a lot of charity work in rural areas, in spiritualist[11] health centres. He was therefore very sensitive to poverty-related problems. In his view, his main duty was to apprise 'the poor' of the difficult responsibilities their desire to have a family entailed. Despite his explicit critique of the 'fascist' birth control campaigns of the past, his words echo the neo-Malthusian discourse of the past. At the same time, however, new topics emerge in his discourse, such as parenthood as an emotional need and the right of 'the poor' to have a family. When asked to explain why he used the term 'poor' in the masculine, the majority of his patients being women, he explained. 'Planning is for couples. Sometimes husbands come because they are sceptical, but here we cannot see them. We do what is doable.'

Dr Fabio's female, 'white'[12] colleagues also emphasized that family planning was a 'couple's right', but argued that the limited space and time allotted for each patient in public health services precluded the possibility of seeing couples. Another structural problem highlighted by most health professionals was the lack of resources: third generation Pills, vaginal rings, hormonal patches and the hormonal IUD were not reimbursed. Moreover, the stock of contraceptives sometimes changed from one month to the next, obliging their patients to continue to use a contraceptive they did not like or to change a contraceptive they wanted to use depending on the method's availability. Finally, most family planning services, with the notable exception of the clinic 'Gine-Salvador', were located at public health centres where gynaecologists trained to insert IUDs were not always available. In this context, it was difficult to meet patients' needs.

However, when their patients complained about their problems with temporary contraceptives, gynaecologists usually blamed their 'ignorance' or their 'absurd beliefs'. 'These patients', or 'these women', most of them used to say, 'trust more their mother or their neighbour than their doctor: this is the real problem'. Their private, upper-middle class patients, they said, also had fears related to medical contraception, but they usually understood instructions and did what the doctors told them to do. Dr Beatriz, a gynaecologist in her 40 s, described her patients at the public family planning centre as follows:

> People here are ignorant, stupid. Sometimes I think I am not clear enough, but then I talk to my colleagues and it's the same. These women don't understand anything[.]. What kind of training do they have in biology? None. They are almost illiterate. Education here is a shame! Massive orientation is given by the TV and on TV there are only soap operas, which are even inciting them. Here women have a lot of pregnancies. Poor women have a lot of children, each one with a different man. I don't know how it is over there [in Europe], but here poor men have a lot of families, a lot of women and a lot of children. And these children will be the thieves of tomorrow, the murderers. (Interview with Dr. Beatriz, Salvador 2004)

The old, neo-Malthusian discourse on family planning as a means to prevent the main social and economic problems of the country, including criminality, clearly emerges in Dr Beatriz' discourse. The solution of these problems was, in her mind, family planning education, which was provided at 'Gine-Salvador' by two auxiliary nurses. Gynaecologists did not have time for that. They spent most of the limited time they had for their patients studying their medical records, almost without looking at their patients in the face, and prescribing routine exams and contraceptives, while distractedly listening to their complaints. When their patients complained about contraceptives' side effects, they usually assumed a paternalistic attitude and explained that their symptoms were 'normal' side effects of the contraceptives they were using.

'Patients often complain about menstrual irregularity and frequent bleeding or amenorrhea' remarked Dr. Fabio one day. 'But these are possible, normal side effects of hormonal injections. If menstruation does not occur, they fear pregnancy. Sometimes they also think the blood can go to their heads...'. After a brief pause he concluded:

> It is part of our culture. Women need more information. Their neighbour arrives and tells them, 'If I were you, I wouldn't use this method any more. It is responsible for all of the bad things happening in your life.' If her husband left her and her son is sick, it is because of a given contraceptive method. Therefore, it is important that female patients talk with their health care providers and confront these myths. (Interview with Dr. Fabio, Salvador 2004)

The main obstacles to family planning were, in Dr Fabio's view, his patients' 'myths' and 'culture', 'our culture', or 'Brazilian mythological thinking'. Brazil, most gynaecologists pointed out, was still backward, or not entirely developed, and their low-income patients, who were originally from 'underdeveloped', rural regions and/or had a low educational level, were not 'enlightened' (*esclarecidos*) enough yet. The main objective of family planning was therefore to 'enlighten' them with the bio-medical rationale. For this reason, according to Dr Celia, a colleague of Dr Fabio with ten years of professional experience at the clinic, 'Gine-Salvador' provided a daily class, aimed at 'explaining the menstrual cycle [...] what the Pill does to hormones and how all contraceptives worked'. Yet the explanations offered during the class were not always 'enlightening'. As the following section will show, many participants left the class with doubts and unanswered questions.

The impossible arithmetic of hormones and the limits of family planning

Around 8:00 a.m. an auxiliary nurse, a 'black'[13] middle aged woman wearing a white, bright uniform, started the class that was mandatory for all new patients. Most of the participants were low-income, 'black'[14] women in their early 20s to early 30s, mothers of one or two children on average, searching for a reliable contraceptive, ideally with no side effects. A few men, the vasectomy candidates, also participated in the class. Karen and Eliete, the teachers, usually started it in the same way, explaining what family planning meant – 'planning how many children one wishes to have and leaving time between pregnancies' – and making a list of all contraceptives available at the clinic. Then they showed the first poster, which concerned the Pill. It depicted a middle class, white woman sitting on a bed, dressed in a blue night shirt, in the middle of taking a tablet. Moonlight shone on the wrappers for the different pills located just below her. Then they explained, in a monotone:

> 'There are three types of Pills: the 21 day Pill, the 28 day Pill or the 35 day one. The last Pill is for breast feeding mothers. Usage of the 21 day Pill: you begin taking it on the first day of your period. After 21 days you must wait seven days while you menstruate. It is normal for your blood to be dark. On the eighth day of your cycle you must start taking the Pills again. You cannot become pregnant during the break. If you don't want to menstruate you should take the Pills right after you finish a packet. Nothing happens. Any questions?' (Fieldnotes, Gine-Salvador, Salvador 2004)

Karen and Eliete always highlighted the importance of following the mathematical formula of 21 days plus a seven day break. However, some women always asked questions such as: 'After the seven-day break do I start using the Pill again whether or not I have had my period?' These questions showed that the Pill's mathematical formula left many women unconvinced. In fact, the way the Pill reacted in the body and its relation with menstruation were not clear to most of them. In their view, conception was the result of the fertilizing activity of the sperm and of the blood contained in the uterus and its product was the ovum,[15] while hormones were associated with sexual desire. Many said, quoting what they had heard Dr Coutinho say on TV,[16] that the male hormone, contained in large quantities in men's bodies, was responsible for an increase in sexual desire, while the female hormone (contained in high quantities in the Pill) was responsible, in their view, for reducing it.

The two auxiliary nurses rarely explained during the class how hormones impacted on ovulation, and hormonal contraceptives' action on it. The participants' educational level was very low, they explained during their interviews, so they tried to use a simple language and to avoid scientific explanations. Eliete talked extensively about ovulation and fertilization only twice in one year. Both times she described the sperm as the protagonist in the fertilization process and the ovum as its passive recipient. Moreover, when she talked about the IUD she made an analogy between the IUD's copper, defined as being able to 'kill the sperm', and the Pill's action.

After one of the two classes I spoke with three young women who had heard Eliete's explanations and asked no questions (fieldnotes, Salvador 2004). They were respectively 25, 20 and 16 years of age. The oldest was the only one who had a stable relationship, while the younger ones were struggling to complete secondary school while working and helping their mothers at home. They were discussing whether to use the hormonal injections 'to enjoy the Carnaval' without worrying, even if they were afraid of gaining weight. When I asked them whether or not 'the Pill inhibits, prevents ovulation' as Eliete had affirmed, the oldest answered, 'The Pill does not prevent ovulation. It protects you. It keeps the sperm from coming into the ovum... It also kills the sperm.' The two other women nodded. So I asked them where they had been told the Pill 'killed' sperm. The youngest told me she had already attended a family planning class while the oldest said, 'The nurse said it during class, didn't she?' Finally, the third woman said that her cousin had told her the same thing. We then talked about the IUD. The copper in IUDs, the three women explained, 'also kills. It destroys.' This destructive action was necessary because, as the youngest woman said, 'The sperm goes into the tubes very, very happily.' 'The ovum does, too, doesn't it?' I asked. 'Yes', she answered, 'but the sperm goes in much quicker.' 'Yes, much quicker', her friends answered in chorus. Based on what they knew before coming to the clinic and on what Eliete had said, the three young women had concluded that the action of the Pill was analogous to that of the copper, which 'killed' sperm.

Emily Martin (1991) has argued that the emphasis on the sperm's active, fertilizing action and the marginalization of the ovum (often described as its passive recipient) in the dominant scientific discourse, reflect and contribute to reproducing the western gendered distinction between strong, active men and weak, passive women. Eliete's rather reasonable, or should we say 'rational' choice – to avoid using scientific terms and discussing 'too difficult' topics – left the door open to dangerous interpretations, and her emphasis on the sperm's fertilizing action had serious, practical consequences. In fact, most female patients had started to use the Pill by themselves, following their friends' and neighbours' suggestions, and some of them took it only when they had sexual intercourse (when the Pill performs its 'destructive' action on the sperm). As a result, they suffered from irregular spotting and some had also experienced unintended pregnancies which they had terminated with a self-administered abortion, using misoprostol bought on the black market. For these reasons they had come to the clinic, where Eliete had confirmed their ideas on the Pill's action. They searched for a more reliable contraceptive, able to effectively 'kill the sperm' (unlike the Pill that they did not trust anymore), a drug that they did not need to take every day and had no side effects.

As Rosa, another patient of 'Gine-Salvador', explained: 'I come home very tired in the evening and I sometimes forget to take it. So, I have blood loss [.] it is difficult to remember to take the Pill every day'. Then she added:

> My cousin used the Pill for a long time and she never had a problem. I think she never forgot to take it. She works in public administration and always has the same daily schedule. After work, she can go back to her own apartment where she can relax because she doesn't have any children. Her whole life is planned. I think that's why she doesn't forget the Pill. If you work morning till night without a fixed schedule and if you come home very tired, it is easy to forget to take the Pill. (Fieldnotes, Gine-Salvador, Salvador 2004)

Rosa was 28 and worked downtown as a housekeeper. She lived in a peripheral neighbourhood and came home very late at night. As with many other family planning users of her age or older than her, she was separated from her first husband and had two children who she had to raise alone. Moreover, she did not have a regular job contract, so not only did she not have any right to maternity leave and paid sick leave, but she could not afford private medical insurance to cover her gynaecological visits, or to buy expensive contraceptives.

Rosa's physician, Dr Celia, said laughing during her interview that she also forgot the Pill when she was young. She abandoned it because, she explained, 'during the weekend I changed routine, so I forgot it'. She therefore switched to hormonal injections, but did not like them, so she tried the IUD, contraceptive implants and she finally chose to use the hormonal IUD. She preferred it because she did not menstruate and she did not suffer from any unpleasant side effects, such as weight gain. Despite having similar problems with the Pill, Dr Celia could try a wide range of contraceptives and did not mind not having regular menstrual periods, while Rosa could only 'choose' hormonal injections or the IUD, but was afraid of the potential negative impact of both on her body and health.

Women's struggles to preserve their health and to involve men in contraception

Most women complained that both the Pill and hormonal injections made them gain weight and provoked headaches. Women using three-monthly hormonal injections also complained they felt 'bloated' (*inchada*), because they did not have regular menstruations. They were very worried because, they said, 'Menstruation is important. You know you are not pregnant' and because 'menstruation is healthy. It cleans you out'. Its suspension was therefore very problematic. Some users of the

Pill also complained that it made them feel 'nervous/anxious' (*nervosa/angustiada*) or 'more nervous' than usual, therefore they had less patience with their children.

The IUD, which could be an alternative to hormonal methods, was defined by most women as an 'external body', which could provoke uterine cancer and other diseases. Some also expressed the fear that physicians would not remove it if they decided for any reason to change contraceptive. In fact, when IUD users complained of heavy menstrual periods, which made them feel 'weak' (*fraca*) and had a negative impact on their work performance (many were domestic workers), physicians always highlighted that this was a 'normal side effect' of this 'excellent long-term contraceptive', 'with no heavy side effects', as most described it, and suggested that they wait the symptoms out.

Instead of responding to their patients' fears and doubts, the physicians' 'clinical gaze' (Good 2003) searched exclusively for the signs of real, true symptoms and diseases, and discarded all the 'false' ones, attributing them to patients' 'ignorance' or 'absurd beliefs'. The two auxiliary nurses also highlighted the 'absurdity' of their patients' 'beliefs' (*crenças*) and practices. One day a man asked during the family planning class if it was possible for a woman to get pregnant while using hormonal injections or the Pill. Karen said that if she was using these methods consistently this was not possible and added: 'The patient must be diligent.' Another man commented then: 'I heard something funny about Coca-Cola. There are women who take douches with it'. Karen commented ironically, 'There is also a thing about salt water. Women think that it washes everything away, the ova too!'. The audience intoned a collective 'Ave Maria!' accompanied by general laughter. The conversation seemed to be over. But a woman spoke up and asked timidly if urine killed sperm. Scandalised, Karen answered 'No, can you imagine?'. The audience again burst out laughing. Some women, however, were not laughing and one of them, a middle aged black woman, said, indicating the men sitting in the class: 'Life is easy for them. Why isn't there a Pill for men?'. Karen answered jokingly, asking the class: 'and you think they would use it?'. All men answered with a loud, collective 'No, for God's sake!' and the conversation finished there.

In the waiting room, many women complained amongst themselves and affirmed during their interviews that sex and contraception were difficult to plan, and saying 'no', 'not now', or 'not without a condom' to a man could provoke tensions and violent discussions (this was why they did not like the diaphragm, as it must be inserted some time before sexual intercourse). Many married women preferred to give their husbands condoms to use 'in the street' with their lovers, and not ask them to use them at home to avoid being suspected of being unfaithful. Most women also said, however, that they did not like condoms, because 'it burns', or 'it interrupts pleasure'.

Karen and Eliete rarely mentioned the male and female condoms because most of their patients did not trust condoms and did not like using them, and also because, as Karen explained, 'the priority here is family planning, not STDs prevention', which was provided by other health services. They also rarely discussed non-medical contraceptives, which were often used by their patients when they stopped using medical contraceptives because of their side effects, or when they 'forgot' to use the condom. Most women used withdrawal and/or a vaginal douche and a glass of salt water to drink at the time of sexual intercourse. 'Salt kills sperm' they explained. 'Doesn't sea water stop menstruation? And isn't sea water salty? It is the same thing'.[17]

Latour (1993) argues that by establishing a distinction between 'belief/culture' and 'knowledge/science' to 'purify' rationality from the polluting influence of prejudice, myth and human passions, modernism has separated what cannot actually be separated: 'Nature', 'Society' and 'Culture'. Bahian women's problems with medical contraception clearly show that 'natural' phenomena, such as (wanted or unwanted) pregnancies, and all their potential or actual consequences (including unsafe abortions) are the result of an ensemble of heterogeneous factors – family planning policies, social and gender inequalities, and cultural values and norms orienting reproductive practices, including those mobilized and (re-) produced via bio-medical discourse.

Challenges for family planning: 'Machismo', STDs and infertility

Most female family planning users considered men's irresponsibility in the sexual and reproductive domain the inevitable result of their 'nature' or of '*machismo*' (male domination), as the youngest ones said. They therefore took care of contraception, or tried to do so, even if they did not like to be charged with the practical and moral responsibility of it. '*Machismo*' was defined by most health professionals, particularly female health professionals, as an important 'cultural barrier' to family planning. Dr Celia, a young gynaecologist in her mid-30 s, defined it as a 'cultural issue', a specifically male attitude preventing men from all social classes from actively participating in contraception, particularly from using condoms. 'They leave it up to us. It's your obligation, your role', she said during her interview. Then she added:

> The number of vasectomies is increasing here, while the number of tubal ligations is dropping. Yes, men participate, they get vasectomies. However, we rarely see a woman accompanied by her husband or boyfriend... here they don't participate jointly in family planning. (Interview with Dr. Celia, Salvador 2004)

Seen from the perspective of the clinic's director, on the contrary, the increase in the rate of vasectomies was a success. 'We had to fight against many dragons', he said referring both to men's diffused fear of impotence and to the hostility of the Catholic Church to his clinic's campaigns aimed at promoting vasectomy. However, he agreed with Dr Celia on one point: the increase in vasectomies and the decrease in tubal ligations were not directly related to each other. Tubal ligations had decreased because women were better educated and had 'internalized' family planning. As Dr Celia explained:

> I think that when a woman becomes aware, enlightened... is educated, she will prefer to use a reversible contraceptive method. She will gain awareness that life can improve and that she can have a better husband and she can be in an environment where having another child is possible. Normally, when you ask to get your tubes tied, it is both because you have a horrible husband and you do not want to have his baby, or because you are not financially able to have another child [.] Many of their husbands are alcoholics. (Interview with Dr. Celia, Salvador 2004)

Like most female health professionals, Dr Celia described her patients' male partners as being irresponsible husbands and fathers, resisting condom use, fearful of vasectomy and often aggressive, if not violent. In her view, the priority of family planning services should be men's involvement in sexual and reproductive health services and STDs' prevention and treatment. Moreover, the government should introduce free treatments for female and male infertility, which were foreseen by the 1996 law on family planning but were never actually implemented. As Dr Celia explained:

> Family planning includes conceiving children when you cannot [.] One of the causes... the main causes of high infertility is STDs, tubal obstruction. If we make information available to prevent this... we should be treating infertility [...] I think that we even have a debt to the population. (Interview with Dr. Celia, Salvador 2004)

An infertility unit had been open at 'Gine-Salvador' for a while. However, the success rate was very low because most patients came for free appointments and free tests but did not have enough money for the medications and treatments. Some also asked to reverse their tubal ligation or vasectomy, but this was impossible because these surgeries were not reimbursed by the SUS. The service was therefore closed.

Reproductive rights and wrongs:[18] Sterilization and abortion

According to Dr Celia and the social worker, the remorse rate was especially high among sterilized men.[19] This happened because, the social worker explained, 'A man has a family. He has a vasectomy performed. Afterwards, when his wife gets older, he leaves his family and starts another. Normally he marries a much younger woman and this woman usually does not have children and wants to start a family'. To prevent later regret 'Gine-Salvador' had established higher age limits for candidates for both male and female sterilization than those established by the law – 30 years of age instead of 25 (or more, depending on the number of living children) – and, as the law establishes, at least two living children (the youngest had to be at least one year old). Moreover, unmarried patients or those in unstable relationships were usually invited to postpone their decision. Exceptions were made only in case of serious health problems.

One day a woman asked Karen if her daughter could be sterilized. She was 29, she explained, and had only one child, but had already had one abortion. Karen responded dryly, 'Abortion doesn't count. With only one child, you can only do it, here, when you are 40 years old.' In her patients' view, on the contrary, the experience or the fear of unwanted pregnancies and unsafe abortions did count and was one of the main reasons why they searched for effective contraceptives, including tubal ligation.

Young women in their early-mid 20 s tried to persuade the social worker and doctors to 'tie' or 'cut' them, but they rarely obtained what they wanted at 'Gine-Salvador'. Unemployment and separations were the reasons most of them mentioned when the social worker asked them why they did not want to have children anymore. Similar reasons were mentioned by most candidates for vasectomy (in their mid-30 s to mid-40 s, on average), who had separated and married again and could not provide for both their families. The social worker always asked male candidates if getting sterilized was their 'free choice'. She did not ask this question of women because, she explained, women knew what they wanted. 'I don't have the means (*não tem condição*)', they used to say, and then explained that they preferred to invest money in their existing children's (private) education than having more children.[20] Out of the medical office they also said that they wanted to have a tubal ligation because 'children are a lot of headache', 'a lot of work' and they could not count on anybody's help at home. Some also added that they wanted to enjoy sex without suffering contraceptives' side effects. According to most women and men interviewed, the ideal number of children a couple should have was two. Those women who had more than two children were labelled as 'irresponsible' and 'shameful', mainly if they were young. To young mothers, sterilization therefore meant at the same time being free of the stigma and fear of 'unplanned' pregnancies and illegal, unsafe abortions, and having more time to enjoy themselves and to complete secondary education, which would allow them to have, they hoped, better jobs and a better future than their mothers.

Conclusions

Family planning emerges from the ethnography presented and discussed in this paper as an ensemble of discourses and medical practices disciplining low-income women's sexual, contraceptive and reproductive life by stigmatizing those who do not embody the neo-Malthusian rationale still orienting many health professionals in their medical practice. At the same time, however, female family planning users 'tactically' use it not only to be good or better mothers and citizens, but also to study, have more leisure time and enjoy sex free from the fear of unwanted pregnancies and unsafe abortions. The bio-politics of the population of the past, aimed at limiting the fertility of the poor and the bio-politics of the 'self' theorised by Rose (2007) seem to coexist in this scenario.

In a paper in which she discusses her ethnography of the use of hormonal treatments in Salvador da Bahia, the anthropologist Emilia Sanabria points out that 'both the regulation of population and the corporealisation of subjectivity can thus be said to operate in Brazil' (Sanabria 2010, 392), the first one in the public health sector and the second one in the private sector. 'The shift from one to the other', she argues, 'can neither be spatialised nor historicised' (Sanabria 2010, 392), as Rose (2007) suggests. This paper shows that these two biopolitical rationalities actually coexist, but it argues that both operate within the public sector. Family planning is in fact experienced by most female family planning users both as a 'duty', a heavy moral and social responsibility they have to carry to avoid stigmatization and to contribute to the well-being of their family, and as a means of personal enhancement. Even health professionals consider and treat their low income female patients both as citizens entitled to rights and self-fulfilment, and as 'irrational', 'irresponsible' women (and men) whose fertility has to be controlled both for their own wellbeing and for the wellbeing of the broader society. The biomedical 'instrumental rationality' (Good 2003) they embody, the lack of time allotted for each patient and the neo-Malthusian rationale still orienting their practice lead them to ignore or consider as irrelevant their patients' psychological, family and social conditions, and to undermine their problems with medical contraception, due to their 'ignorance' and 'culture'.

'Culture' was conceptualized by most health professionals as an ensemble of 'backward' beliefs and 'traditional', 'absurd', 'irrational' practices that patients would abandon after being 'enlightened' by the biomedical rationale. Instead, the female participants in the family planning class often left it with a lot of doubts and unanswered questions about medical contraceptives' action in their bodies. This was the result of health professionals' choice to avoid using scientific terms and explanations with their 'ignorant' patients – a 'rational', or at least reasonable choice, in their perspective – and of their authoritarian and patronising attitude.

The interaction between health professionals and patients was shaped by what Fassin defined a 'biopolitics of otherness' (Fassin 2001). This biopolitics operates by labelling and stigmatizing the Other/s (immigrants in France in the case discussed by Fassin; low-income, black Bahian women in the case discussed in this paper) as 'different' because of their naturalized, essentialized 'culture... considered as a hereditary characteristic of the individual' (Fassin 2001, 7), while recognizing them, at the same time, as entitled to rights and citizenship by virtue of their status of 'sufferers'. This excluding inclusion, as Fassin argues, neutralizes the possibility of considering and effectively addressing social inequalities and discrimination. A similar mechanism operates in the family planning services where this study was undertaken. Female family planning users were in fact described by most health professionals as both victims – of poverty, lack of access to good health care and education, and, finally, *machismo* – and as 'irresponsible', bad mothers and/or 'irrational', bad patients stuck in their 'culture' and 'absurd beliefs'.

Gender inequalities were highlighted both by health professionals and their patients but they were rarely discussed openly. Most health professionals thought that women should take control of contraception, because men, and particularly their low-income patients' partners, were 'irresponsible' partners and fathers. In this way they reconfirmed the traditional gendered distinction between 'irresponsible' men, led by '*machismo*' to prioritize their own desires in the sexual and reproductive domain regardless of their partner's and children's desires and needs, and 'responsible' women, who are expected to take care of both reproduction and contraception.

Most of their female patients had partially internalized this '*machista*' moral distinction between 'responsible' women and 'irresponsible' men. At the same time, however, they complained about having to assume entirely the practical and moral responsibility of contraception (and therefore of contraceptive failure, unwanted pregnancies and abortion) and called for more participation of men and more research on male contraceptives. They did not passively accept being treated as 'irresponsible' mothers and 'bad', 'irrational' patients. They appropriated and creatively manipulated the fragments of 'scientific' discourse circulating in the media and at the clinic, mixing it with what they had learnt at home. By condensing 'multiple and often conflicting social and semantic domains' (Good 2003, 172) in order to produce a meaningful interpretation of contraceptives' action in their bodies, these women created what Byron Good would define a new 'semantic network' (Good 2003). Moreover, they resisted in practice the scientific, objectified interpretation of their "normal" symptoms by abandoning certain contraceptives and searching for better ones. Sometimes, they also rejected the 'authoritative knowledge' (Davis-Floyd and Sargent 1997) health professionals embodied, by using non-medical contraceptives. This decision, however, often exposed them to unintended pregnancies and unsafe abortions and left them with few alternatives, as public health services did not always provide a lot of choice and they did not have enough money to buy their favourite contraceptive. In this context, tubal ligation appeared to many of them as the only or the best solution to all their problems, and when they could not obtain it in the public health system, they looked for a local politician providing free surgeries in private clinics in exchange for a promise of a vote.

If reproductive freedom means 'that both women and men have viable alternatives from which to choose, and that the best social and political conditions exist that allow women to decide, free from coercion or violence, if, when and how many children to have' (Lopez 2008, 143), Bahian family planning users, particularly women, are not actually 'free'. However, they are not only and simply 'docile bodies' (Foucault 1975) on and through which biopower operates. They struggle to preserve their health and pursue pleasure and use family planning services and contraceptives as 'tactics' (De Certeau 1990) to realize their multiple, conflicting desires: on the one hand, they wish to enjoy sex while avoiding unintended pregnancies, unsafe abortions, and the side effects of medical contraceptives; on the other hand, they strive to be recognised as good, responsible citizens with social and moral duties, who also have rights, and seek recognition of the right not only to good quality family planning services, but also to better jobs and social services supporting the reconciliation of work and family life.

Acknowledgements

The author is grateful to Milena Marchesi, co-editor and co-organizer of the 2008 American Anthropological Association session on which this paper is based, for constructive feedback and discussions. The author also wishes to acknowledge the stimulating comments of the discussant for the paper, Dr Elizabeth Krause, and of two anonymous reviewers. Funding for the dissertation research on which this paper is based was made possible by a Refeb grant (Exchange Program between France and Brazil, financed by the French Embassy in Brazil). Ethical approval for this research was not required by the EHESS (Ecole des Hautes Etudes of Sciences Sociales) of Paris, France, where the author obtained her PhD. However, the research was authorized by the Directors of both clinics where it was undertaken, oral consent was obtained prior to each interview with health professionals and patients and anonymity was granted to all interviewees.

Conflict of interest: none.

Notes

1. All translations from Portuguese into English are by the author.
2. In the mid-1960s fertility rates started to decline in the wealthiest regions of Southern Brazil and among upper-middle classes. In the following decade they dropped in all regions and classes (see Bozon and Enoch 1999).
3. Despite being very close to the Catholic Church and strongly pro-natalist, the Military government assumed an ambiguous *laissez-faire* attitude towards family planning organizations. For the history of family planning in Brazil see Fonseca Sobrinho (1993).
4. The author undertook her PhD ethnographic research during two periods of six months each in 2003–2004 and a final month in 2005 in two family planning services and in a peripheral neighbourhood in Salvador da Bahia. She engaged in participant observation two to three half-days a week at a clinic and a maternity-hospital and conducted unstructured interviews with 11 health professionals and 70 family planning users, mostly

low-income black women. At 'Gine-Salvador' clinic 27 female patients, 10 male patients, five physicians, two auxiliary nurses and one social worker were interviewed.

5. 'Black' translates here both '*preto*' (black) and '*pardo*' (brown, mixed), two of the terms used in the Brazilian national census to define colour/race, along with *branco* (white), *amarelo* (yellow) and *indigena* (indigenous).
6. The name of the clinic is fictitious.
7. In the early 2000s, the administration of family planning services was being transferred from the State to local Councils and many public health centres were undergoing structural reforms, therefore family planning services were not regularly provided.
8. After two caesarean sections, any pregnancy and delivery presents higher risks, therefore sterilization can be medically justified. On the use and abuse of caesarean sections in Brazil see: Berquó (1993).
9. On the scientific and political debate on women's sterilization and reproductive rights in Brazil see Berquó (1999); Côrrea and Avila (2003).
10. All personal names in this paper are fictitious.
11. Introduced at the end of the nineteenth century from Europe (France) into Brazil, Spiritualism is nowadays practised only by a minority of Brazilians – 1.3% in 2000. The spiritualist movement, however, is very concerned with education, science and charity and owns nurseries, health centres and hospitals, so many health professionals are affiliated to it. Most health professionals interviewed by the author declared themselves as non-practising catholic, while most female patients interviewed did not regularly frequent any Church or '*terreiro*' (Afro-Brazilian religions' place of worship). Only a minority regularly frequented evangelical Churches. In 2000, most Brazilians declared themselves as catholic, but the Catholic Church lost followers in recent decades (from 89.2% in 1980 to 73.8% in 2000), while evangelical Churches multiplied their followers (from 6.6% in 1980 to 15.4% in 2000), and the rate of non-religious Brazilians increased from 1.6% in 1980 to 7.3% in 2000 (IBGE 2000).
12. Dr Fabio's colleagues defined themselves as '*brancas*', white, while he defined himself as '*pardo*', literally light brown. Most health professionals working at family planning centres were women.
13. Both auxiliary nurses defined themselves as '*negras*', which can be translated as 'black' in British English. This term, which was denigratory in the past, was re-introduced by the Black Movement in the last decades to assert racial pride and black power.
14. The English term 'black' is used here to translate both *morenas* and *pardas* ('brown', 'light brown' or 'mixed race' in English) the two terms most female family planning users adopted to define their colour, and *negras* ('black') which was mainly used by the youngest ones.
15. Similar findings have been found by the Brazilian anthropologist Ondina Fachel Leal in southern Brazil (see Leal 1995).
16. On Coutinho's theories on hormones see Coutinho (1996, 1988a) and Manica (2009).
17. On the contraceptive use of salted water in Salvador da Bahia see Rabelo (2001).
18. Reference is made here to the book by Betsy Hartmann (1995), '*Reproductive rights and wrongs: The global politics of population control*'. See references.
19. No reliable data on sterilized women's and men's regret after sterilization were available at the clinic.
20. Similar arguments were used by the women interviewed by Dalsgaard in a poor neighbourhood of the northeastern capital of Recife (see Dalsgaard 2004).

References

BEMFAM (Sociedade civil bem estar familiar no Brasil). 1986, 1996. *Pesquisa Nacional Sobre Demografia e Saúde*. Rio de Janeiro: Macro Int. Inc.

Berquó, E. 1993. Brasil, um caso exemplar? Anticoncepção e partos cirúrgicos à espera de uma ação exemplar. *Revista de estudos feministas* 2, no. 2: 366–81.

Berquó, E. 1999. Ainda a questão da esterilização feminina no Brasil. In *Questões da saúde reprodutiva*, ed. K. Giffin and S. Costa Hawker, 113–26. Rio de Janeiro: Ed. FIOCRUZ.

Bozon, M., and E. Enoch. 1999. Brésil: la transition démographique d'un pays hétérogène. *Population & Sociétés* 345: 1–3.

Caetano, A.J., and J.E. Potter. 2004. Politics and female sterilization in northeastern Brazil. *Population and Development Review* 30, no. 1: 79–108.

Côrrea, S., and M.B. Avila. 2003. Direitos sexuais e reprodutivos. Pauta global e percursos brasileiros. In *Sexo e vida*, ed. E. Berquó, 17–78. São Paulo: Unicamp.

Coutinho, E. 1996. *Menstruação. A sangria inútil. Como não menstruar sem remédio. Uma analise da contribuição da menstruação para as dores e os sofrimentos da mulher*. São Paulo: Ed. Gente.

Coutinho, E. 1998a. *O sexo do ciúme*. Salvador: Memorial das Letras.

Coutinho, E. (1998b). O Nordeste e o Problema demográfico Brasilerio. In *O descontrole da natalidade no Brasil*, 73–83. Salvador: Memorial das Letras.

Dalsgaard, A.L. 2004. *Matters of life and longing. Female sterilization in Northeast Brazil.* Copenhagen: Museum Tusculanum Press.

Davis-Floyd, R., and C. Sargent. 1997. *Childbirth and authoritative knowledge. Cross cultural perspectives*. Berkeley, CA: University of California Press.

De Certeau, M. 1990. *L'invention du quotidien. 1. Art de faire*. Paris: Gallimard.

Diniz, D., and M. Medeiros. 2010. Aborto no Brasil. Uma pesquisa domiciliar com técnica de urna. *Ciência e Saúde Coletiva* 15, no. 1: 959–66.

Fassin, D. 1996. *L'espace politique de la santé*. Paris: Puf.

Fassin, D. 2000. *Les enjeux politiques de la santé*. Paris: Karthala.

Fassin, D. 2001. The biopolitics of otherness. *Anthropology Today* 17, no. 1: 3–7.

Fonseca Sobrinho, D. 1993. *Estado e população: uma historia do planejamento familiar no Brasil*. Rio de Janeiro: Rosa dos Tempos.

Foucault, M. 1975. *Surveiller et punir. Naissance de la prison*. Paris: Gallimard.

Foucault, M. 1976. *Histoire de la sexualité 1. La Volonté de savoir*. Paris: Gallimard.

Good, B. 2003. *Medicine, rationality and experience*. Cambridge: Cambridge University Press.

Hartmann, B. 1995. *Reproductive rights and wrongs: The global politics of population control.* Boston, MA: South End Press.

Heilborn, M.L., M. Bozon, E.M.L. Aquino, and D. Riva Knauth. 2006. *O aprendizado da sexualidade. Reprodução e trajetórias sociais de jovens brasileiros*. Rio de Janeiro: Garamond and Fiocruz.

IBGE. 2000. National 2000 census, data on religion-Brazil: http://www.ibge.gov.br/home/estatistica/populacao/censo2000/populacao/religiao_Censo2000.pdf

Latour, B. 1993. *We have never been modern*. Cambridge, MA: Harvard University Press.

Leal, O. 1995. Sangue, fertilidade e praticas conceptivas. *Corpo e significado*. 13–36. Porto Alegre: Ed. da Universidade.

Lopez, I. 2008. *Matters of choice. Puerto Rican women's struggle for reproductive freedom*. New Brunswick, NJ: Rutgers University Press.

Manica, D. 2009. Imperativos da natureza: sexualidade, genero e hormonios na produção de Elsimar Coutinho. In *Sexualidade, Reprodução e Saúde*, ed. M.L. Heilborn, E.M. Aquino, R.M. Barbosa, F.I. Bastos, E. Berquó and F. Rohden, 261–84. Rio de Janeiro: FGV Editora.

Martin, E. 1991. The egg and the sperm. *How science has constructed a romance based on stereotypical male-female roles. Signs* 13, no. 3: 485–501.

Menezes, G., and E.M.L. Aquino. 2009. Pesquisa sobre aborto no Brasil: avanços e desafios para o campo da saúde coletiva. *Cadernos de saúde pública* 25, no. Sup 2: 193–204.

Ministerio da saúde 2008. *Pesquisa Nacional Sobre Demografia e Saúde da Criança e da Mulher 2006*. Brasilia: Ministerio da saúde.

Morgan, L.M., and E.F.S. Roberts. 2009. Rights and reproduction in Latin America. *Anthropology News* March: 15–16.

Rabelo, M. 2001. *Da água com sal à laqueadura. Um estudo etnográfico com mulheres sobre práticas contraceptivas. Dissertação de Mestrado*. Salvador: ISC-UFBA.

Rose, N. 2007. *The politics of life itself. Biomedicine, power, and subjectivity in the twenty-first century*. Princeton, NJ: Princeton University Press.

Sanabria, E. 2010. From sub- to super-citizenship: Sex hormones and the body politic in Brazil. *Ethnos* 75: 4, 377–401. DOI: 10.1080/00141844.2010.544393/ URL: http://dx.doi.org/10.1080/00141844.2010.544393.

The right to have a family: 'legal trafficking of children', adoption and birth control in Brazil

Andrea Cardarello

Université du Québec en Outaouais, Canada

This paper focuses on one of the 'child-trafficking scandals' that occurred in Brazil in the 1990s. Ethnographic research was carried out between 2000 and 2001 within a movement of poor families formed in São Paulo to put pressure on the authorities to review the legal procedures that had led to their children being placed for national and international adoption. Fieldwork was supplemented by other data, including reports by legislative bodies, articles in the press, and case files involving the termination of parental rights. This paper explores views on international adoption among members of the Brazilian elites such as judges, agents in the field of child protection and journalists, in the context of old but persistent neo-Malthusian ideas. Although the Brazilian birth rate is now below the replacement level, it is still common to blame 'irresponsible' reproduction among the urban poor for violence in large cities. Drawing a parallel with the routine sterilization of women that prevailed for decades and was encouraged by Brazilian physicians, the paper examines how, in a 'struggle against poverty', judicial agents took it upon themselves to enforce 'birth control' through adoption, bypassing family consent and the law in the process. The paper concludes by arguing that discrimination against poor families who are viewed as disorganized, immoral and irresponsible – characteristics frequently associated with criminality by a sector of the elites – has contributed to the view that lower-class families do not have the right to bear children, or to keep them.

In an article on adoption published in the national magazine *Isto É* in, 1995, the federal police acknowledged the existence of 'legal child trafficking' in Brazil, explaining that they could not intervene once a court adoption decree was produced 'except when the decree was shown to be manifestly illegal' (*Isto É* 22 February 1995, p. 80). This official use of the term 'legal child trafficking' confirms the findings of Abreu (2002, 47–8), that illegality in the adoption process in Brazil was widespread in the courts during the 1980s and 1990s with the support, complicity, and even active participation of judges and other agents of the judicial system. These irregularities, however, although common in cases of domestic adoption, only become unacceptable to Brazilian society when they benefit foreigners. Thus, while the term 'irregular adoption' is used when Brazilians adopt children within an illegal but sometimes legitimized framework, that of 'trafficking' is used when children are adopted by foreigners.

The subject of international adoption is at the root of an inflammatory discourse regarding the Brazilian nation (Abreu 2002, 152–3). On one hand, certain people in the judicial milieu and the elite are ashamed of the country's poverty. They see international adoption as a solution, as it allows poor children to be raised by people living in rich countries, where they would benefit from 'every comfort' and the best available health care. On the other hand, judicial 'nationalists' and a number of journalists took a stand against international adoption, emphasizing that it was shameful that Brazil did not have the means to raise its children and had to give them up to foreign couples. Through this discourse, international adoption became associated with trafficking and child exporting (Abreu 2002, 88). The Child and Adolescent Act of 1990 gave rise to a discourse that advocated the child's right to remain within the family and in the community of origin (and, therefore, in the birth country), strengthening arguments against international adoption. The rights of poor families to keep their children are rarely mentioned in this nationalistic debate, and neither are the ways in which these children are taken and placed not only in foreign families, but also in wealthy Brazilian families.

Based on ethnographic research carried out between 2000 and 2001 in the city of Itaguaí[1] in the State of São Paulo, this paper focuses on one of the 'child-trafficking scandals' linked with international adoption that occurred in Brazil in the end of the 1990s. In 1998, on the initiative of a lawyer, about 20 poor families formed a movement and demanded a review of the judicial procedures that resulted in their children's adoptions. This paper draws on fieldwork carried out by the author among the families participating in this movement. Thirty members of 20 families were interviewed on the cases of 30 adopted children. Moreover, 15 lawyers and social workers who had contact with the families were also interviewed. Finally, two reports by federal and state legislative bodies, over 100 articles published in the press, nine files of cases of termination of parental rights, printed interviews and other written testimonies were also examined.

The first part of this paper clarifies the procedures of the Itaguaí case of legal child trafficking. The second part draws a parallel between the arguments of judges and other actors in the judicial domain who promote adoption and those of the doctors who encourage the sterilization of poor women. The third section examines how the concept of 'disorganized families' is used to disqualify poor people from raising children. Finally, the last section explores the persistence of neo-Malthusian ideas in the Brazilian public agenda. The paper concludes by arguing that the question of who has a right to have a family in Brazil is strongly influenced by the perception of poor families as disorganized, immoral, and irresponsible – characteristics that are frequently associated with criminality.

'Irregularities' in the Itaguaí case

Between 1992 and 1998, poor families from the town of Itaguaí lost approximately 480 children to adoption, with close to 200 of them being adopted internationally. A single judge and a single state prosecutor authorized all these adoptions.

At the end of the 1990s, the Human Rights Commission of the Legislative Assembly of São Paulo and a commission of parliamentary inquiry of the Federal

Senate produced two reports denouncing irregularities in the adoption procedures of the children of the families concerned. According to these reports, in more than one case, allegations of child abuse and abandonment – the basis for the court process that led to the termination of parental rights – had never been proved. Irregularities observed in other cases of child-trafficking scandals in Brazil have been corroborated, such as adoptions processed within very short time periods; suspected collusion between judges and attorneys; non-existent files or suspicious dates and signatures in files.[2]

Moreover, after the children had been removed from their families, several family members went to the court to search for them or to obtain information. Although the courts thus had personal contact information for these families, they did not send a bailiff to summon the families to defend themselves, but instead published a public notice summons (*edital*), as if the families' addresses were unknown. These summonses were issued in a legal gazette, the *Diário oficial*, to which the families did not have access. Thus, in several cases, the legal process of the termination of parental rights began and ended without the families even being aware of it.[3] When families were aware that the process of the termination of parental rights had begun, the court refused to provide the families with any information whatsoever; they were simply advised to find a lawyer. The cost of a private lawyer is prohibitive for these poor families, and no one at the court informed the parents that they were eligible for legal aid.

Some of the families had signed documents without knowing their content. Either they were illiterate – some could sign their names but not read – or they did not understand the legal terms used in the document. They were also frequently coerced to sign by the commissioners for minors and the social workers, sometimes immediately after their babies' births, when mothers and babies were still in hospital. Some said they had been tricked: they had signed declarations without reading them (saying they had 'put their faith in the legal system'), under the impression they were signing a document authorizing visiting rights or the return of their children. Only later were they informed that they had signed a renunciation of their parental rights. When the families protested against the removal of their children, the commissioners for minors often threatened to put them in prison. Because they refused to sign, some parents were actually held at the police station for several hours.[4]

The creation of the Movement of families and the acquittal of the judge

At the end of 1998, the print media reported on the presence of a group of about 100 families who gathered regularly in the square in front of Itaguaí's courthouse. These families demanded a review of the judicial procedures that had resulted in their children's adoptions. Its name had been suggested by a lawyer, inspired by the Argentinean movement *Madres de la Plaza de Mayo* (The Mothers of the Plaza de Mayo), the association became known as the 'Movement of the Mothers of the Courthouse Square' (see Cardarello 2007, 2009). Two years later, each vigil was attended in its entirety by only seven people, on average. Despite the small number of core participants remaining, extended family continued to appear at the gatherings, if only to catch up on the latest news.[5]

The judge and the State prosecutor in the Itaguaí case founded an NGO that claimed to protect children and deprived families. Journalists' and politicians' suspicions of child trafficking were aroused when it was discovered that donations for the organization came from an Italian association that promoted international adoption by Italian couples.[6] The Senate inquiry revealed no irregularities in the judge's bank accounts, but a number of checks made out to the State prosecutor by the same Italian association were uncovered.

In December, 1998, due to the pressures and denunciations, the judge and the prosecutor in question were transferred to courts in the city of São Paulo – courts without jurisdiction over the youth and child sector. However, despite the inquiries that were carried out, the creation of the movement by the families of origin and the ensuing media attention, the judge responsible was acquitted of all accusations of irregularities by the Court of Justice of São Paulo at the end of 2001.

The heavy media coverage of the 'child-trafficking scandals', such as the Itaguaí case, and the series of regional investigations that followed led to a significant decline in the number of international adoptions in the country in recent years.[7] Brazil is no longer among the top ten sending countries as it was from 1980 to 1998 (being fourth on the list in the 1980s; Selman 2009). However, instances of child-trafficking in national and international adoption continue to be reported by the press.[8]

Abandonment and circulation of children

Some of the Itaguai children placed for adoption by the judicial system were living with grandparents or other extended family members, while others were living with *famílias de criação* (foster families) who were not biologically related to them. In Brazil, it is not uncommon for children to circulate between different family milieus (the homes of grandmothers, other relatives, close friends, or neighbors), a pluriparentality practice that makes several adults responsible for the child's socialization.[9] These children may speak of having two or more mothers, even when their biological mothers have maintained a relationship with them. In fact, some of the Itaguai children had been using the word 'mother' for biological mothers, aunts and extended family members who were raising them before their removal from their families.

One aspect that stood out in the analysis of the stripping of parental rights in the Itaguaí cases was that children who were not living with their parents were labeled 'abandoned' by the judicial authorities. Thus, the child-rearing practice of 'child circulation' became the equivalent of child abandonment in the case files.

In one case file, a commissioner for minors reported that a mother had left her child in the care of a female friend since birth. According to the judge's decision, this situation was treated as abandonment, as the mother had 'abandoned the child to an unauthorized stranger'. As stated by the legal authorities, leaving a child in the care of 'an unauthorized third person' indicated 'immoral' behavior on the part of the parent(s). According to the judge, only the father or the mother had the right to the custody of the child; *criação* parents 'could not be considered family members' – because family implies blood relationships. The arguments presented in the case files by the judge and the state prosecutor made it clear that formal adoption was seen as the only way to guarantee a life of 'stability and dignity' for the child, and that the fact that the child was staying with different people at different times implied an 'inability to establish a relation of affinity and security, or to recognize parental figures'. As the families had not passed through the courts to legally formalize the custody rights or the adoption of a child and had arranged informally to have the child brought up 'in circulation', the child was therefore seen to be in an 'irregular situation' and to be living with the *familia de criação* in an 'illegal manner'.

'Having the means' (*ter condições*)

Another salient aspect in the analysis of the files and the interviews with the families was that the families in these cases were accused of 'not having the means' (*não ter condições*). In Portuguese, this expression can be used very vaguely. Unlike English, it is not necessary to specify the means for what – such as meeting the children's material needs, or providing an adequate moral environment or a minimal living space (with a bedroom for the children, for example). Nonetheless, when the expression 'you do not have the means' is used without further qualifications, it generally refers to the lack of financial resources, and this is what the families understood it to mean.

According to many sources, such as interviews with families and social workers, the legislative reports and articles in the press, after the removal of their children family members were told in court they had to fulfill certain conditions such as getting a job or moving to a dwelling with more than one room. They were then misled into believing that if they complied, their children would be returned to them. In addition, unrealistic deadlines (a few days, in some cases) to meet these conditions were set. Moreover, some of the better-off members of these families went to the court to declare their willingness to take care of the children. However, this fact was never recorded and was absent from the case files.

It should be recalled that Article 23 of the Child and Adolescent Act of 1990 made important changes to previous legislation with respect to the relationship between adoption and poverty.[10] Under Article 23, 'the lack of material resources does not constitute a sufficient motive for the loss or suspension of parental rights', and, if this is the only motive, 'the child or adolescent will remain in the family of origin, which will be referred to official aid programs'. Since the law was adopted, several legal practitioners in favor of international adoption have protested against this article (Abreu 2002, 29). They evoke the 'nationalization of poverty', arguing that parents who are too poor to meet their children's basic needs should lose their parental rights.

The interpretation of Article 23 by the magistrate in his decisions was quite peculiar. The magistrate's view was that the lack of resources was 'always' or 'generally', associated with other motives that justified the termination of parental rights. However, several justifications for the termination of parental rights cited in the judgments were directly related to the 'lack of material resources'. If the parents were not able to rent a house or apartment, or pay for their children's medical treatment or a daycare centre, the decision was made to remove that child from his or her family. Citing Article 23, the magistrate emphasized that this measure was not at all taken due to the parents' poverty, but to guarantee the child's fundamental rights. To take on the characteristics of an accusation – and because of the existence of Article 23 – the arguments in the judgments use moral and psychological categories such as 'irresponsibility', 'immaturity', 'moral shortcoming' or 'lack of the emotional structure needed for motherhood'. Thus, in the files, the ubiquitous 'lack of material resources' takes the form of a 'violation of the children's basic rights', such as the right to good health, the right to a 'structured life', the right to 'a responsible and effective family', or the right to a 'decent, dignified life'. Therefore, parents' rights were violated under the pretext of protecting their children. In order to gain a better understanding of the procedures of the judicial authorities in this case, it is necessary to take into consideration the importance of neo-Malthusian ideas in the Brazilian public arena.[11]

A 'culture of sterilization'

Neo-Malthusian ideas emerged on the Brazilian political scene during the 1960s, when the Kennedy administration tried to link its economic aid to Latin America with policies aimed at checking the so-called population explosion (Barroso and Bruschini 1991, 153–4). In American politics after the Cuban Revolution in 1959, there was the fear that Communist doctrine could make the prevailing poverty of Latin America a factor encouraging the rise of leftist governments. As Brazil was viewed as a major player in world overpopulation, a political menace was thus transformed into a demographic one (Pedro 2004, 32–7; see also Martine 1998, 183).

In 1965, the International Planned Parenthood Federation (IPPF), an international agency largely funded by the United States government, created a private institution, BENFAM, which provided free family-planning services through community pill distribution and clinics. BEMFAM's educational program aimed to convince poor people that the reason for their poverty was the large number of children they had, while women were pictured as irresponsible baby-producers (Barroso and Bruschini 1991, 153–4). Finding support among the conservative elite, birth control was viewed as the solution to the problem of poverty. However, in the face of opposition from the Catholic Church, leftists, and certain members of the military dictatorship, the Brazilian government, even if it tolerated the presence of private organizations such as BEMFAM, showed no willingness to implement an official program to curb population growth.[12]

Even without an aggressive birth-control strategy by the government, the Brazilian total fertility rate has declined since 1970, from 5.6 to 1.83 in 2007 (IPEA 2008). There is not a simple and ready explanation for Brazil's accentuated fertility decline; rather the phenomenon can be explained by a multiplicity of factors.[13] This process occurred at different times in the different regions of Brazil. In the south and centre-west, the drop in the fertility rate began in the early 1970s, while in the north and northeast, it began at the beginning of the 1980s, with the 'unofficial' sterilization policy discussed below (IBGE 2009a, 10).

According to Martine (1998, 175), two methods of birth control have had the most impact on fertility reduction: abortion and sterilization.[14] Regarding female sterilization, even if its rate had diminished since 1996, when it was the method adopted by 40% of women living with a partner, sterilization remains the dominant form of contraception in the country. The 2006 PNDS report found that 30% of women living with a partner had been sterilized, followed by 21% who used contraceptive pills.

The Brazilian demographer Elza Berquó (1998) has stated that the promotion of tubal ligation as the main contraceptive method has led to the establishment of a 'culture of sterilization' in the country. Before 1997, sterilization was not officially performed by the public health service, although doctors and their female patients found ways to do it daily, most often by combining a C-section with a tubal ligation (Diniz, de Mello e Souza, and Portella 1998, 36).[15]

Although the uncontrolled practice of sterilization and its effects have been thoroughly criticized in Brazil by women's groups and feminist politicians, it is still an extremely common practice among the poor (Dalsgaard 2004, 28). According to the 2006 PNDS report, the percentage of sterilized women with four years of education or less was more than double that of women with nine years or more (PNDS 2006, 79). Moreover, the percentage of sterilized women in the North and Northeast, the poorest regions, is more than twice that of women in the South and Southeast of Brazil. The rate is also much higher among black women than white women.

The reasons why poor women in Brazil are more likely to use sterilization or abortion rather than non-permanent methods to control their fertility are many and complex. In a study among low-income Brazilian women in an outlying neighborhood of the northeastern capital of Recife, Dalsgaard (2004, 127) reported the lack of availability, the unreliability and uneven quality of pills and condoms, a lack of knowledge about alternatives, and the lack of power to negotiate in sexual relations as being among the reasons that women did not find any reversible contraceptive method satisfactory. Like other researchers, Dalsgaard cites unequal gender relations – in frustrated marital relationships, for example – as another reason for lower-income women to want to control fertility (Dalsgaard 2004, 30, 205, see also Goldani 2002).

Along with structural problems and gender inequality, the disdainful attitudes and behavior of medical providers towards low-income patients emerge in several studies as another significant factor influencing poor women in their reproductive and contraceptive decisions. In clinics, sterilization is often suggested to them as the only desirable contraceptive method (Diniz, de Mello et Souza, and Portella 1998, 59).[16] Medical providers tend to negate women's role as decision-makers, especially when they are poor and black, and to promote methods that they consider 'more efficient' for such women.

In a study undertaken in the Northeast, in the state of Pernambuco, Caetano and Potter (2004, 81–2) focus on the common practice promoted by local politicians, physicians and politicians–physicians of the exchange of free female sterilizations for promises of votes. The authors assert that, aside from women's growing demand for birth control, physicians played a direct role in the Brazilian fertility rate decline by the medicalization resulting from the expansion of hospital care.

Caetano and Potter mention monetary profit, electoral success, and compassion or charity among the motives for those who provide sterilizations. However, in the Brazilian context, the persistence of 1960s neo-Malthusian ideas and the concept of poverty as inseparable from criminality can be added to these motives. Dalsgaard (2004, 115), for example, quotes the statement of a gynecologist who was known to provide sterilizations and who was also a local politician:

> Some of the women are not so sure, some even regret it later, but the majority really needs [to do the sterilization] as they do not have the conditions [(or means) to take care of more children]. If I could decide, I would do them all. I do not think that tubal ligation is good, but I think it is bad to suffer from hunger. At times I send food because they are dying from hunger, they bring the small children and I give cans of milk, because the small children are dying from hunger and they have nothing at home, and I feel sorry for them. To ligate tubes isn't good, but they do not know how to avoid more

> children. Isn't it better to ligate than abort, suffer from hunger, and bring up *marginais*?[17]

Based on a sample of 400 interviews with women in a low-income community also in Pernambuco, Recife, Fernandes (2003, 257) concluded that the high number of sterilizations among the women in the sample – almost half of whom had been sterilized – was not necessarily a consequence of the women's decisions, or their full consciousness of what the decision implied. The desire of many of these women to have more children was hidden and substituted by the discourse of their doctors: 'The doctor said I already had too many children'. As in the pro-adoption discourse, the term 'lack of means' takes centre stage again. Many women justified their decision to undergo sterilization by the doctors' urgings, stating 'the doctor said I didn't have the means'. The great majority of the women, however, did not know exactly to which means the doctors were referring.[18]

Dangerous classes

The position of the judges, lawyers, journalists, politicians, agents in the field of child protection and other members of the elite who promote international adoption parallels that of the doctors who advocate the sterilization of poor women. In their eyes, adoption not only prevents these children from becoming 'miserably poor' or 'hungry', it also prevents them from becoming 'criminals'. In one of the first interviews granted to the press following the founding of the Movement of families in Itaguaí, the judge stated: 'I'm an unconditional fan of international adoption. It's the way to get rid of prostitutes and drug dealers'. Ginzburg (1990, 91), quotes a statement by one Brazilian judge who encourages international adoption:

> Of course, in the adoption of a child, one must proceed with great caution, respecting the laws of both the country of origin and the country where the child will live. Nevertheless, put yourselves in our place: the future for these children is often the street, miserable poverty, and in the best-case scenario, prison.

The association between poverty and criminality with respect to poor children has been made since the beginning of the twentieth century, when the question of childhood and childhood criminality first emerged in Brazilian society (see Rizzini 1995). Since then, there has been a continual process of shifting back and forth between protecting the child and protecting the society that is potentially threatened by that child.

Itaguaí's inhabitants, the conservative and progressive wings of the Catholic Church, and members of the press were divided on the procedures of the judicial authorities. However, articles supporting the Court decisions (written not only by journalists but also by attorneys and other authorities) showed the approval of a sector of Brazilian society. One journalist asserts in a *Jornal da Cidade* article on 9 August, 1998:

> That judge acted according to the law. He removed children from inconsequential and irresponsible parents – children who would certainly have gone from begging at intersections (when still small enough) to petty larceny (puberty), to armed robbery (as young adults), and to today's typical vices, like marijuana, cocaine, heroin, crack, and other less common substances (for life) and end up in a bad situation: probably in a reformatory, a FEBEM [a public shelter for young people], or in an adult prison.

In the view of another columnist, the judge had saved the children from living in 'pigsties'. In the media's support of the judge's position, terms like 'dirty', 'irresponsible', and 'scatter-brained' were used to describe the families in the Movement.

'Disorganized families'

Many changes have occurred in Brazilian families over the last decades. Since the 1970s, there has been a decrease in the number of families headed by couples. With the increasing participation of women in the workplace, the number of single-parent families headed by women has increased. In Brazil, this rate rose from 15.8% in 1996 to 18.1% in 2006 (IPEA 2008). The importance of legal marriage has diminished. As in other parts of the world, new reproductive technologies have created diverse family arrangements and, after overcoming resistance from some quarters, it is now possible for homosexual couples to legally adopt children (see Fonseca and Cardarello 2010; Scott 2004).

The family was always viewed by politicians, doctors, and law makers as the locus of morality and the ideal milieu for children's socialization, in preference to institutions or the street (Londoño 1991; Rizzini 1993; Alvim and Valladares 1988). However, not all families are considered adequate to the task of child-rearing. Whenever a discussion of the problem of childhood poverty in the country arises, the family of origin is often blamed for the child's situation, being described as 'disorganized' (*desorganizada*), 'maladjusted' (*desajustada*), or 'disunited' (*desagregada*).

The 'disorganized family' is defined by Brazilian child welfare agents and policy-makers as a family model 'characterized by the absence of the father and in general headed by the mother' (Alvim and Valladares 1988, 19; Pereira Júnior, Bezerra and Heringer 1992).[19] In a study carried out in the mid-1990s in a shelter for children and adolescents in the city of Porto Alegre in southern Brazil, the psychologists interviewed said that, in their view, a 'disorganized family' was one in which there were 'undefined roles, instability, switching mothers and fathers all the time' (see Cardarello 2000). Other employees of the institution added the parents' lack of effort to organize themselves, which basically implied getting a steady job, providing an adequate dwelling, and ensuring that the children were properly nourished and attended school. The 'organization' category as linked to the notion of hygiene was also used by social workers when describing the condition of the dwelling at the time of their visits (an 'organized home' was a clean and tidy home).

These same connotations related to the term 'disorganized family' or its synonyms (as 'maladjusted' family) were seen in the files documenting the termination of parental rights in Itaguaí. In one case, three sisters aged 6, 11, and 12 years were reported to a commissioner for minors by a policeman because they had left home without permission and gone to the centre of town to beg. According to their maternal grandmother, only the youngest was detained, as the two older girls managed to flee. However, the next day, when the older sisters went to try to obtain their little sister's release, they too were held by the commissioners for minors and were placed in an institution.

In this legal procedure, the family's functioning was described as 'maladjusted'. The father of the three girls, of a family of nine children, was working in agriculture in another state. In one of the hearings to determine whether the children should be removed from the family by the judicial authorities, the judges and the public prosecutors made accusations that the father was 'completely absent in the raising of the children' and that the mother was 'idle'. The father explained that the mother looked after the children when he was away working. According to the lawyer who defended the family, at that time, the father wanted to separate from the mother, a fact that was also used against the parents. The lawyer advised him to wait, as a parental separation would make the case more difficult to win.

In another case, according to a mother who was interviewed, the magistrate asked her at one of the hearings if she owned a house or had a husband. Her lawyer objected that owning a house or being married did not constitute a legal condition for a mother to keep her child and that he himself was not a home-owner. The magistrate reacted by treating the lawyer with disdain, telling him that he 'knew nothing about family matters'.

The analysis of the procedures in the termination of parental rights shows that the use of the organization/disorganization opposition and other pairs of contrastive terms – such as 'well-adjusted/maladjusted' – was promoting the family model that the agents find adequate for a child's development. As others authors suggested (Alvim and Valladares 1988, 19; Pereira Júnior, Bezerra, and Heringer 1992), this model is the bourgeois nuclear family, in which the parents work, occupy a dwelling, and possess the necessary 'means' or 'conditions' for them to be able to meet their children's needs.

Neo-Malthusian ideas: a national debate

The following statements were overheard in an anteroom before a meeting of the town's custodial council (*Conselho Tutelar*) of the city of Itaguaí, an organization responsible for ensuring observance of the rights of the child and adolescent.[20] The president of the council, speaking to an employee, said:

> The judge probably made some money out of it, that's the problem [*esse é o mal*]. But he did 250 good deeds.[21] These women who have no jobs and no husbands, horrible dwellings and mistreated children, and who are without means... There are lots of people who can look after children, who can offer them another kind of life. [Otherwise], what is a child going to be like 12 years from now? [...] There's proof that the judge has an apartment across from the courthouse [bought with the money he made]. But in my opinion, that's better than leaving children in the street [...]. If the children were sold, it's because there are people who can pay. People without any money shouldn't have children. [...] You're going to look after your child, not the municipality, the State, or the shelter. I wanted to have one child. You have to think of the future—school—, you have to guide them, it's not easy. [...] I'm not ready to have another child [...] And the Church against using condoms – what's that?'

According to Pedro (2004, 29–30), the fall in the birth rate since the middle of the twentieth century in both 'developed' and 'poor' countries produced a very diversified discursive struggle. In Europe in the 1960s and 1970s, for example, the availability of the birth control pill was viewed as a gain for women's rights. However, in Latin America, it was qualified as a necessary measure to counter the menacing 'population explosion'. The arguments to justify family planning policies in poor countries arose within power struggles in which class prejudice, the confrontational context of the Cold War, and racism all had significant weight.

According to the IBGE (Brazilian Institute of Geography and Statistics), in spite of all the changes in reproductive behavior that have been observed in the country, major sectors of society are still stuck in an ideological framework founded on the idea that the country will continue to experience stages of a process previously referred to as a population explosion (IBGE 2009a, 5n6). It is very common to read in newspaper articles or letters to the editor that the cause of the rise in violence in Brazil in the last two decades was the absence of family planning.

In 2007, the declarations of the Governor of the State of Rio de Janeiro promoting the legalization of abortion as a way of combating crime led to the publication of polemics in the press. In the daily *Folha de São Paulo* on October 27, sociologist and demographer George Martine and anthropologist Sônia Corrêa, in a letter printed to contrast with statements by an economist, argued that the politician's remarks reflected the simplistic reasoning among many that poor people have too many children, which generates more poverty, which in turn generates more violence. They also wrote that in spite of the Governor's good intentions, his view covered up the real causes of complex social problems like violence by scapegoating the poor, mainly women. The year before, Martine, as president of ABEP (*Associação Brasileira de Estudos Populacionais* – Brazilian Association of Population Studies), wrote to the weekly news magazine *Veja* contesting the content of an article on the population question (Martine 2006). The article cited lepidopterist Paul Ehrlich, who said in an interview: 'over-population is our key problem'. Martine's letter maintained that in Brazil 'the ready acceptance of the idea that poverty, crime, environmental destruction, and other evils can be blamed on demographic pressure is scandalous'. Even though his letter was supported by the signatures of dozens of specialists, the magazine decided not to publish it.

Conclusion: The right to have a family

In the files that were examined documenting the termination of parental rights, the phrases used in connection with poor families – 'lack of dignified and honest work', 'lack of family structure', and 'precarious sanitary conditions' – convey in legal or bureaucratic language what the president of the Custodial Council, among others, expressed in a simpler way: 'these parents do not have jobs'; 'these mothers do not have husbands'; and 'these families are dirty'.

The legal cases discussed in this paper are good examples of what Bourdieu (1987) calls 'the universalising posture of the law', when the juridical sphere contributes to imposing a representation of normality in which any different practice is seen as deviant, anomalous, or even pathological. Ignoring the plurality of family forms existing in the different social classes of Brazil, the law reflects only one possibility of family arrangements: the dominant model of the nuclear family. The informal practices of the circulation of children and pluriparentality are either not taken into account or condemned, and the points of view of the families of origin are not considered.

Nevertheless, these cases can be seen as simply an example of class prejudice. To a significant portion of Brazilian society, poverty is a moral failure. In the opinion of the journalists, lawyers, and politicians who supported the judicial authorities, 'legal child trafficking' is tolerated because these families have no right to reproduce due to their immorality, laziness, uncleanliness and irresponsibility – characteristics considered endemic to their social class.

As Reis shows (2005, 42–4), the Brazilian elites' concept of poverty is characterized by an apparent dissociation between social awareness and social responsibility. These groups do not acknowledge the interdependency that exists between them and the bottom layers of the society, nor are they convinced of the need for collective social solutions. Although Brazilian elites may acknowledge that the poor are victims of major social forces, to these privileged sectors of society, the urgent reasons to fight poverty and inequality are the danger and the threat of violence in large cities. Therefore, from that perspective, they see the urban poor as a 'dangerous class'. In the Itaguaí case, the discourse of the judge and his supporters is representative of the contemporary version of the elite's old fear of poor children, who they see as presenting a threat in the present (street children), or the future (poor children will become 'criminal' youths and adults, or 'prostitutes').

The liberal bourgeois concept that predominates in the legal procedures that were examined is that of a family able to support itself, living in a dwelling in which parents carry out the duty of helping their children, who are not obliged to work (Pereira Júnior, Bezerra, and Heringer 1992). The connotations that are still attached to the term 'disorganized family' in the child protection milieu do not take into account the huge economic and cultural gap that separates poor families from state agents. Concerning issues of entitlement to parenthood and 'stratified reproduction' (Colen 1995, 98), the Itaguaí case reveals who, in Brazil, has the right to have a family (see Abreu 2002, 49; Fernandes 2003, 254).[22]

Even with a total fertility rate of 1.8 children, a rate that is below minimum population replacement level, neo-Malthusian ideas persist in the Brazilian public arena. In a peculiar anti-poverty strategy, similar to that of physicians who advocate the sterilization of poor women in Brazil, certain legal agents take it upon themselves to enforce birth control among the lower classes. Through legal child-trafficking, they authorize adoptions without respecting the legal formalities and above all, without the consent of the families of origin. This process reflects the idea, among the elite, that poor families do not have the right to bear children, or to keep them.

Acknowledgements

This research project was made possible by a grant from CAPES (*Fundação Coordenação de Aperfeiçoamento de Pessoal de Nível Superior*) and by financial assistance from INRS – *Urbanisation, Culture et Société*, following the expiration of the Brazilian grant. The author is grateful to Darcy Dunton for the translation of the first version of this paper, as well as Silvia de Zordo, Milena Marchesi, Betsy Krause, Joshua Moses, Cristiano Martello and two anonymous reviewers for their help and comments. This research was approved by the Ethical Committee of the Université de Montréal.

Conflicts of interest: none.

Notes

1. Itaguaí is a fictitious name.
2. See Abreu (2002) for the same irregularities in cases in the state of Ceará.
3. For a similar use of public notice summons to search for parents in some Argentinean courts nowadays, see Villalta (2008).
4. One mother told me that a commissioner for minors, who was also a policeman, pointed a gun at her head when four of her five children were removed.
5. The ethnography collected suggests that, due to the consequences of the sudden removal of their children, many other families, having received no support, were not even able to join the movement. For some parents and relatives, the removal of their children caused the manifestation or aggravation of problems of alcoholism and mental illness, and even suicide attempts.
6. Between 1994 and 1998, families from Italy adopted about 40% of the Brazilian children placed for international adoption (Commissione per le Adozioni Internazionali, http://www.commissioneadozioni.it/stat/1.htm; and Fonseca, 2002). In 2004, with an official total of 477 international adoptions, Brazil still ranked fifth among the countries of origin for international adoption in Italy – and tenth in France (Selman 2009, 37). In the case of this 'scandal', the press also reported adoptions by German, Dutch, Swiss, American, and Danish families.
7. Nevertheless, statistics for domestic adoptions are not available. For other countries in which the 'child trafficking scandals' caused a drop in the number of international adoptions, see Selman (2000, 24).
8. To give a few examples, in 2009, there were media allegations of the abuse of power, concealment of information, and the arbitrary removal of the children of at least 42 families for purposes of adoption in the State of São Paulo between 2004 and 2007, involving custodial councilors (*conselheiros tutelares*) and a state prosecutor (see Constantino 2009). In 2006, in the State of Parana, the police arrested six people for child-trafficking. Of the six, three were lawyers, and one was a Civil Police Force support agent. They had been involved in the sale of children to couples all over Brazil (Agência 2006). In 2003, international adoptions in Pernambuco were investigated for suspicion of child trafficking, due to fast-track judicial processes in which foreigners were willing to pay $7500 for a smoothly-processed adoption (Associação 2003). The media report pointed at public employees of the Civil Registry Office and the administrative department of the Juvenile Court who allegedly adulterated documents in cases that were processed in only three days.
9. A number of kinship studies carried out among the lower classes in Brazil have brought out the importance of this practice (see Fonseca 2002; Sarti 2003 [1996]; Cardoso 1984).
10. Even with the considerable progress made in the seven years of the Lula government (2003–2010), with official statistics revealing that 20 million people rose above the poverty line out of a total population of 190 million, Brazil still has one of highest rates of income disparity in the world (see Rocha 2010; Barros, Henriques, and Mendonça 2001). It should be noted that poverty affects Brazil's black population more than its white population (see Guimarães 2002).
11. For the adoption of the principle of the 'child's best interest' contributing to the family's being frequently considered the primary cause of the violation of children's rights, without any responsibility being attributed to the State or of the society in general, see Cardarello (2012). In 2009, the 'Child and Adolescent Act' of 1990 was amended, to guarantee the child's right to live in the family of origin. However, it is too early to appreciate the impacts of this change.
12. In the Catholic church, the rejection of family planning policies is a point of consensus in a church otherwise divided with regard to political and doctrinal issues. Although no longer linked to the state, Catholicism is still the dominant religion in the country, as it is the majority religion, including among the elite (Barroso and Bruschini 1991, 154).
13. For an analysis of this factors see Martine (1998).
14. Abortion is illegal in Brazil except in cases of rape or endangerment of the woman's life. Most abortions are obtained illegally in the private sector. Abortions range from the

safest type of operation in private clinics for members of the middle and upper class who are able to pay to procedures done at home using traditional methods and remedies (see Diniz, de Mello et Souza, and Portella 1998, 36).

15. It should be noted that 43% of deliveries in Brazil are by Caesarian (IBGEb 2009b, 11).
16. For a description of attitudes of neglect and the brutality of members of hospital staff and the feelings of shame and inferiority that these attitudes caused in lower-class women see Dalsgaard (2004, 149–55). For a case of a woman sterilized against her will in São Paulo, see Nelson (2002, 204).
17. The word *marginais* here, as the author explains in a footnote (Dalsgaard 2004, 222, n22), usually refers to criminals and drug users in popular language.
18. It is interesting that in Dalsgaard's study about the reasons why some women so willingly chose to be sterilized in a poor neighborhood close to Recife, in the Northeast region, a frequent explanation was: 'we don't have the conditions for bringing up children' (Dalsgaard 2004, 20). If having no children was out of question, having many was seen as irresponsible behavior (Dalsgaard 2004, 16). Viewing these women as simultaneously controlled and empowered, the author advocates that recourse to sterilization represents a search for social acknowledgement, as well as a hope for control over their lives (Dalsgaard 2004, 98, 25–7). And yet, as the author asserts, in Foucauldian terms, these women had incorporated their ascribed role as responsible citizens and, simultaneously, as docile bodies, through their identification with the role of compliant patient (Dalsgaard 2004, 135).
19. This concept corresponds to the dysfunctional family category used in the United States from the 1960 s to describe poor African-American families (Pine 1996, 227), and, in the 1970 s, to the irregular families category in France (Meyer 1977). See Meyer (1977) and Donzelot (2005 [1977]) for this categorizing of poor families as a form of discipline in the Foucauldian sense.
20. The local custodial councilors register children and families in social, educational, and financial-assistance programs, exercise control over associations dealing with children and adolescents, and place children in institutions when they deem it necessary.
21. In fact, as mentioned above, it is estimated that approximately 500 adoptions were authorized by the judge, the state prosecutor, and his assistants.
22. For the concept of 'stratified reproduction' used to describe the power relations by which some categories of people are empowered to nurture and reproduce, while others are disempowered, see Colen (1995) and Ginsburg and Rapp (1995, 3).

References

Abreu, D. 2002. *No Bico da Cegonha. Histórias de Adoção Internacional no Brasil.* Rio de Janeiro: Relume Dumará.

Agência de Notícias Estado do Paraná. 2006. Polícia prende seis pessoas acusadas de tráfico de crianças. August 4, http://www.aen.pr.gov.br/modules/noticias/article.php?storyid=21998&tit=Policia-prende-seis-pessoas-acusadas-de-trafico-de-criancas (accessed March 3, 2007).

Alvim, R., and L. Valladares. 1988. Infância e sociedade no Brasil: Uma análise da literatura. *BIB* 26: 3–37.

Associação Brasileira de Magistrados e Promotores de Justiça da Infância e Juventude. 2003. Adoções em Timbaúba-PE estão sob suspeita de tráfico. Clipping, *Diário de Pernambuco*. September 8, http://www.abmp.org.br/noticias.php?origem=2&idt=2585 (accessed November 23, 2006).

Barros, R.P., R. Henriques, and R. Mendonça. 2001. A estabilidade inaceitável: Desigualdade e pobreza no Brasil, *IPEA*, Text for discussion No. 800, Rio de Janeiro, June.

Barroso, C., and C. Bruschini. 1991. Building politics from personal lives: Discussions on sexuality among poor women in Brazil. In *Third world women and the politics of feminism*, ed. C.T. Mohanty, A. Russo and L. Torres, 153–72. Bloomington: Indiana University Press.

Berquó, E. 1998. The reproductive health of Brazilian women during the 'lost decade'. In *Reproductive change in India and Brazil*, ed. G. Martine, M. Das Gupta and L.C. Chen, 380–404. Delhi: Oxford University Press.

Bourdieu, P. 1987. The force of law: Toward a sociology of the juridical field. *Hastings Law Journal* 38, no. 5: 814–53.

Caetano, A.J., and J.E. Potter. 2004. Politics and female sterilization in northeast Brazil. *Population and Development Review* 30, no. 1: 79–108.

Cardarello, A. 2000. Du mineur à l'enfant-citoyen: Droits des enfants et droits des familles au Brésil. *Lien social et politiques* 44: 155–66.

Cardarello, A. 2007. Trafic légal' d'enfants: La formation d'un mouvement de familles pauvres contre les politiques de l'adoption au Brésil. PhD diss., Université de Montréal.

Cardarello, A. 2009. The Movement of the Mothers of the Courthouse Square: 'Legal child traffiquing', adoption and poverty in Brazil. *Journal of Latin American and Caribbean Anthropology* 14, no. 1: 140–61.

Cardarello, A. 2012. O interesse da criança e o interesse das elites: 'Escândalos de tráfico de crianças', adoção e paternidade no Brasil. *Scripta nova*, Revista Electrónica de Geografía y Ciencias Sociales. Barcelona: Universidad de Barcelona, March 15, Vol. XVI, no. 395(10). http://www.ub.es/geocrit/sn-395/sn-395-10.htm

Cardoso, R.C.L. 1984. Creating kinship: The fostering of children in favela families in Brazil. In *Kinship ideology and practice in Latin America*, ed. R.T. Smith, 196–203. Chapel Hill: University of North Carolina Press.

Colen, S. 1995. 'Like a mother to them': Stratified reproduction and West Indian childcare workers and employers in New York. In *Conceiving the new world order: The global politics of reproduction*, ed. F.D. Ginsburg and R. Rapp, 78–102. Berkeley: University of California Press.

Constantino, J. 2009. Tribunal Popular ouve famílias que tiveram filhos levados por Conselho Tutelar. *Brasil de Fato*, October 10, http://www.brasildefato.com.br/node/1588 (accessed October 8, 2009).

Dalsgaard, A.L. 2004. *Matters of life and longing. Female sterilization in northeast Brazil.* Copenhagen: Museum Tusculanum Press, University of Copenhagen.

Diniz, S.G., C. de Mello e Souza, and A.P. Portella. 1998. Not like our mothers – Reproductive choice and the emergency of citizenship among Brazilian rural workers, domestic workers and housewives. In *Negotiating reproductive rights, women's perspectives across countries and cultures*, ed. R.P. Petchesky and K. Judd, 31–68. London: Zed Books.

Donzelot, Jacques. 2005 [1977]. *La Police des Familles.* Paris: Les Éditions de Minuit.

Fernandes, M.F.M. 2003. Mulher, família e reprodução: Um estudo de caso sobre o planejamento familiar em periferia do Recife, Pernambuco, Brasil. *Cadernos de Saúde Pública* 19, no. 2: 253–61.

Fonseca, C. 2002. Inequality near and far: Adoption as seen from the Brazilian favelas. *Law and Society Review* 36, no. 2: 397–431.

Fonseca, C., and A. Cardarello. 2010. Família e parentesco. In *Sociologia: ensino médio*, ed. A.C. Moraes, 209–230. Brasília: Ministério da Educação.

Ginzburg, N. 1990. *Serena Cruz o la vera giustizia.* Torino: Einaudi.

Ginsburg, F.D. and R. Rapp, eds. 1995. *Conceiving the new world order: The global politics of reproduction*, Introduction, 1–17. Berkeley: University of California Press.

Goldani, A.M. 2002. What will happen to Brazilian fertility? Completing the fertility transition. ESA/P/WP.172/Rev.1. *UN Population Division*, pp. 358–75.

Guimarães, A. 2002. *Classes, Raças, Democracia.* São Paulo: Editora 34.

IBGE (Brazilian Institute of Geography and Statistics). 2009a. http://www.ibge.gov.br/home/estatistica/populacao/indic_sociosaude/2009/comentarios.shtm. Accessed December 2010.

IBGE. 2009b. http://www.ibge.gov.br/home/estatistica/populacao/indic_sociosaude/2009/comentarios.shtm. Accessed December 2010.

IPEA. 2008. PNAD 2007: Primeiras análises (Demografia e Gênero). Comunicado da Presidência no 11, Brasília.

Londoño, F.T. 1991. A origem do conceito menor. In *História da Criança no Brasil*, ed. M. Del Priore, 129–45. São Paulo: Contexto.

Martine, G. 1998. Brazil's fertility decline, 1965–95: A fresh look at key factors. In *Reproductive change in India and Brazil*, ed. G. Martine, M. Das Gupta and L.C. Chen, 169–207. Delhi: Oxford University Press.

Martine, G. 2006. Mensagem George Martine a Redação da Veja 12/02/2006. http://www.abep.org.br/usuario/GerenciaNavegacao.php?texto_id=2954. Accessed December 2010.

Meyer, P. 1977. *L'Enfant et la Raison d'État*. Paris: Éditions du Seuil.

Nelson, S. 2002. Constructing and negotiating gender in woman's police stations in Brazil. In *Rereading women in Latin America and the Caribbean – The political economy of gender*, ed. J. Abbassi and S. Lutjens. Maryland: Rowman & Littlefield.

Pedro, J.M. 2004. Entre a bomba populacional e o direito das mulheres. In *Genealogias do Silêncio: feminismo e gênero*, ed. C.S.M. Rial and M.J.F. Toneli, 29–39. Santa Catarina: Ed. Mulheres.

Pereira Junior, A., L. Bezerra, and R. Heringer, eds. *Os Impasses da Cidadania. Infância e adolescênia no Brasil.* Rio de Janeiro: Ibase.

Pine, F. 1996. Family. In *Encyclopedia of social and cultural anthropology*, ed. A. Barnard and J. Spencer. London: Routledge.

PNDS. 2006. Pesquisa Nacional de Demografia e Saúde da Criança e da Mulher: dimensões do processo reprodutivo e da saúde da criança/Ministério da Saúde, Centro Brasileiro de Análise e Planejamento. Brasília.

Reis, E. 2005. Perceptions of poverty and inequality among Brazilian elites. In *Elite Perceptions & Inequality*, ed. E. Reis and M. Moore, 26–56. London: Zed Books.

Rizzini, I. 1993. *Assistência à Infância no Brasil. Uma análise de sua construção*. Rio de Janeiro: Editora Universitária Santa Úrsula.

Rizzini, I. 1995. Crianças e menores, do pátrio poder ao pátrio-dever. In *A Arte de Governar Crianças: História das políticas sociais, da legislação e da assistência à infância no Brasil*, ed. F. Pilotti and I. Rizzini. Rio de Janeiro: Instituto Interamericano del Niño.

Rocha, G.M. 2010. Quel bilan social pour 'Lula'? *Le Monde Diplomatique* 113, no. Oct–Nov: 94–7.

Sarti, C. 2003 [1996]. *A família como espelho: um estudo sobre a moral dos pobres*. 2a. edição revista. São Paulo: Autores Associados.

Scott, P. 2004. Família, gênero e poder no Brasil no século XX. BIB. *Revista Brasileira de Informação Bibliográfica em Ciências Sociais* 58, no. 1: 29–78.

Selman, P. 2000. The demographic history of intercountry adoption. In *Intercountry adoption; developments, trends and perspectives*, ed. P. Selman, 15–39. London: British Agencies for Adoption & Fostering.

Selman, P. 2009. The movement of children for international adoption: Developments and trends in receiving states and states of origin, 1998–2004. In *International adoption, global inequalities and the circulation of children*, ed. D. Marre and L. Briggs, 32–51. New York and London: New York University Press.

Villalta, C. 2008. Entre reformas: procedimientos y facultades en torno a la adopción legal de niños. Actas del IX Congreso Argentino de Antropología Social, Universidad Nacional de Misiones, Argentina.

Reproductive governance in Latin America

Lynn M. Morgan[a] and Elizabeth F.S. Roberts[b]

[a]*Department of Sociology and Anthropology, Mount Holyoke College, South Hadley, MA, USA;* [b]*Department of Anthropology, University of Michigan, Ann Arbor, MI, USA*

This paper develops the concept of reproductive governance as an analytic tool for tracing the shifting political rationalities of population and reproduction. As advanced here, the concept of reproductive governance refers to the mechanisms through which different historical configurations of actors – such as state, religious, and international financial institutions, NGOs, and social movements – use legislative controls, economic inducements, moral injunctions, direct coercion, and ethical incitements to produce, monitor, and control reproductive behaviours and population practices. Examples are drawn from Latin America, where reproductive governance is undergoing a dramatic transformation as public policy conversations are coalescing around new moral regimes and rights-based actors through debates about abortion, emergency contraception, sterilisation, migration, and assisted reproductive technologies. Reproductive discourses are increasingly framed through morality and contestations over 'rights', where rights-bearing citizens are pitted against each other in claiming reproductive, sexual, indigenous, and natural rights, as well as the 'right to life' of the unborn. The concept of reproductive governance can be applied to other settings in order to understand shifting political rationalities within the domain of reproduction.

Introduction

This paper offers the concept of 'reproductive governance' as an analytic tool for tracing the shifting political rationalities directed towards reproduction. The framework of reproductive governance is developed in reference to Latin America, where since the mid-1990s, within the context of neoliberal economic reforms, there has been a barrage of constitutional, civil, juridical and legislative initiatives to both liberalise and curtail reproductive and sexual behaviour through new moral regimes and rights claims (Alvarez 1998; Morgan and Roberts 2009, 78; Corrales 2010). These shifts in reproductive governance provide a framework for understanding what is marked as rational and irrational reproduction in contemporary Latin America, with the hope that the perspective may be usefully applied elsewhere as well.

In recent years, Latin American progressive activists working with international donors have brought about policy changes in the name of gender, reproductive, and sexual rights. Throughout the region, activists have passed legislation to punish domestic violence. They have lobbied legislators to improve access to abortion, emergency contraception, sterilisation, and assisted reproductive technologies and they have supported the rights of sexual minorities, including the decriminalisation of homosexuality and legalisation of gay marriage. At the same time, abortion has been banned entirely in Nicaragua (2006), El Salvador (1998), the Dominican Republic (2009), and several Mexican states, and in Costa Rica, the Constitutional Chamber of the Supreme Court banned in-vitro fertilisation in 2000 to appease the Catholic Church. These efforts have been led in the name of the rights of the unborn, where the foetus is cast as a rights-bearing citizen. By contrast, immigrants may be defined as people who drain the state of resources, as those whose rights can be denied or withheld. Such duelling rights claims produce new kinds of actors and subject positions and new moral regimes.

Moral regimes refer to the privileged standards of morality that are used to govern intimate behaviours, ethical judgements, and their public manifestations. The concept builds on Michel Foucault's 'regimes of truth', the historically specific mechanisms that produce the ideas that function as true (Foucault 1990). Additionally moral regimes incorporate Didier Fassin's notion of the 'politics of life', which refers not only to how populations are governed 'but to the evaluation of human beings and the meaning of their existence' (Fassin 2007, 500–1). With moral regimes of reproduction, the focus becomes the evaluation of actions and ideologies related to generation, perpetuation, and human continuity. Moral regimes are often evaluated in relation to other, supposedly immoral and irrational activities. The specific fault lines along which moral regimes are policed will vary from place to place, but examples may include diverse sexual behaviours and identities, family formations (including marriage, adoption, and inheritance), domestic organisation (including the gendered division of labour), manifestations of religious and spiritual commitments (including the justifiability of secular or theocratic states), and idealised forms of social reproduction (such as education or social security). Post war population control policies, for example, often produce images that serve as ideological guideposts for what modern families should become (Aramburú 1994).

In Latin America, progressives and conservatives alike have made reproduction central to moral regimes that emphasise values of compassion, conscience, nurturance, responsibility, dignity, and citizenship. When Brazilian poverty is discursively associated with criminality, for example, moral regimes may be more supportive of the notion that children of poor families be sent abroad for adoption (Cardarello 2012). Nicaraguan migrants who give birth in Costa Rican hospitals may be more likely than their Costa Rican counterparts to become candidates for tubal ligation, because of their perceived irrational reproduction, at a time when Costa Rican pundits inflame nationalist and pronatalist sentiments by making dire predictions about the consequences of Costa Rican fertility decline and the coming 'demographic winter' (Goldade 2007; Bashford and Levine 2010). In both of these examples, reproductive practices that might otherwise be morally suspect, like tubal ligation within a Catholic nation, are recast as defensible and even rational; this theme is echoed by the other papers in this volume. The activities of the biological body – especially the reproductive and sexual body – are at the centre of these regimes, and are hence critical sites of contention.

Rivalry over rights

Shifts in reproductive governance are facilitated by the now international political legitimacy ascribed to the concept of universal human rights. As the human rights strategy garners favour in international law and diplomacy, an ever-expanding coterie of new constituents (including indigenous peoples, youth, environmentalists, and the disabled, sick, and landless) has begun to frame their social struggles in terms of rights (Goodale and Merry 2007). But for all its rhetorical appeal, the proliferation of rights-talk can be treacherous, because it increasingly allows the claims of rights-bearing citizens to be pitted against one another. This is consistent with David Harvey's observation that the proliferation of rights-based social action, made through judicial processes that tend to favour the well-resourced, is produced within a neoliberal template that promotes civil society at the expense of the state apparatus (Harvey 2005, 78).

Part of the genius of rights is that anyone can use them to make claims on behalf of any individuals as separate from the state. Thus, Latin American conservatives who objected to the language of universal human rights at the Cairo and Beijing conferences in 1994 and 1995, for example, can now be heard appropriating the rhetoric of rights to argue for natural (divine) rights, family (parental) rights, and the 'right to life' of the unborn. The rivalry over rights has led activists on opposite sides of an issue to insert preferential language into a proliferation of duelling laws and international accords; disputes over how to interpret the resulting stalemates are regularly heard by the Inter-American Commission on Human Rights, an intra-hemispheric supra-state institution produced to arbitrate human rights violations. In contemporary juridical and legislative processes, advocates on any side of an issue may cast their opponents' position as immoral, frivolous, profane, un-democratic, or seditious by making claims about the kinds of rights they are ignoring. These arguments about rights constitute new kinds of actors and pit them against older subject positions, for example embryos versus women, subjecting both actors to new forms of reproductive governance.

Reproductive governance refers to the mechanisms through which different historical configurations of actors – such as state institutions, churches, donor agencies, and non-governmental organisations (NGOs) – use legislative controls, economic inducements, moral injunctions, direct coercion, and ethical incitements to produce, monitor and control reproductive behaviours and practices. Along with moral regimes, the concept of reproductive governance is shaped by Foucault, who distinguished between governance through sovereign power (or explicit force) and governance through biopower, in which subjects come to govern themselves on intimate bodily levels (Foucault 1990). According to Foucault, sex was one of *the* exemplary sites for the deployment of biopower in modern European nation states. Biological sex linked anatamo-politics – the disciplining of individual bodies – with biopolitics, the large-scale production and management of populations. 'Sex was a means of access both to the life of the body and the life of the species. It was employed as a standard for the disciplines and as a basis of regulation' (Foucault 1990, 146). Sex is related to reproduction, of course, yet in an era when sex and reproduction have become more separable, they can also be analysed as distinct domains. Within the logic of biology, sex is no longer completely reproductive, nor is reproductive practice necessarily sexual. The field of what now constitutes reproductive practice has thus simultaneously shrunk and expanded. And indeed,

Foucault himself argued that realms other than sexuality – race, for example – can profitably be analysed through the framework of biopower (Stoler 1995; Foucault 2003; McWhorter 2004).

The concept of reproductive governance allows for the examination of how the subject making powers of moral regimes directed towards reproductive behaviours and practices are fully entangled with political economic processes. As historians and feminist social theorists have repeatedly demonstrated, these linkages have been evident since the nineteenth century, when the production and management of populations and reproduction began to be understood as a distinct domain (French and Bliss 2007). Reproduction, which has been made to appear domestic, intimate, and apolitical is fully enmeshed within the production of entities like nation-states and economies (Donzelot 1979; Gibson-Graham 1996; Wilson 2004). The concept of reproductive governance highlights the intersections of international policies – pertaining to migration, health, and reproduction, for example – and those occurring on the national scale where their effects are often executed, experienced, and analysed. Tracking reproductive governance across boundaries helps to illuminate how broader political economic processes – such as the expansion of women's paid formal labour and the incursion of privatised medicine – are shaping reproductive and sexual rationalities and ideologies and making new subject positions (Bedford 2009; Mooney 2009; Ewig 2010). In Latin America, neo-liberal economic processes and related struggles over who shall be worthy of rights, in regards to reproduction, invoke old and new categories and actors like 'indigenous women', 'victims of domestic violence', 'responsible mothers', 'resource depleting migrants', and 'the innocent unborn', who appear to arise on their own as independent entities.

The constitution of such subject positions occurs among international financial institutions, a proliferation of NGOs across the political spectrum, endogenous social movements, and the Catholic Church, as well as newly powerful evangelical Christian churches. Reproductive governance provides a theoretical framework for understanding the regulation of reproductive options, behaviours, and identities available to women and men, who are often constituted as citizens responsible for reproducing rational social and national bodies (Ginsburg and Rapp 1995). Rational citizens are defined as those that embody and reproduce state-supported priorities in their values, conduct, and comportment. The concept of reproductive governance allows for the consideration of the links between embodied and biological moral regimes, national political strategies, and global economic logics, therefore linking 'intimate governance to world governance' (Bashford 2006).

The remainder of this paper illustrates the conceptual utility of reproductive governance to Latin America with a brief history of major trends in population and reproduction from the Cold War to the present. From there the authors deploy the concept of reproductive governance to make three related claims. First, struggles over rights seek juridical codification (that is, constitutional and legislative status) in ways that earlier population approaches did not. Second, contemporary struggles over reproductive governance highlight the tension between individuals and collectivities as bearers of rights. This is consistent with neoliberal efforts to undermine protectionist states, privatise markets, and emphasise the rights of consumers to freedom of choice. Third, concerns about the reproduction of national populations are frequently expressed as control over intra-population movements, especially cross-border migration. Whereas state institutions were once primarily concerned with controlling the fertility of individual citizens, today state institutions are also likely to direct their attentions towards monitoring the fertility and movement of migrants (Krause and Marchesi 2007).

From Cold War overpopulation to the rhetoric of rights

During the Cold War, reproductive policies were directed at controlling population growth and achieving political stability (Hartmann 1999). Reproductive rhetoric was focused on overpopulation, and policies were governed through technocratic regulatory mechanisms and vertical population control campaigns supported by international donors, directed towards bounded nation states. Even without much evidence that Latin America as a whole had a 'population problem' (Aramburú 1994), international donors and state institutions promoted fertility control as a way to improve health, achieve national economic security, public health, hemispheric stability, modernisation, and full industrialisation. Less explicit agendas involved the promotion of smaller families as a way to combat the spread of communism, increase consumption of consumer goods, and reduce specific racial populations. Aid agencies, and state administrations, and even occasionally the Catholic Church, engaged in concerted efforts to make modern contraceptive methods available throughout the region (Carranza 2007; Necochea López 2008).

The end of the Cold War coincided with a period of rapidly declining fertility rates across Latin America as women began to use modern contraception and surgical sterilisation (Leite 2004). Birth rates fell even in nations without explicit population programmes and policies. While the total fertility rate in Latin America had once exceeded the global average, in the late 1980s it fell below the world fertility rate. Shortly thereafter, the rhetoric of reproductive governance shifted from population control to reproductive and sexual health and rights, as public health experts recommended that attention be shifted to programmes that would prevent sexually transmitted diseases, provide abortion services, and treat infertility (Lane 1994).

The Cairo International Conference on Population and Development in 1994 and the Fourth World Conference on Women in Beijing in 1995 hammered out programmes of action based on a rights-based approach to reproductive health (Haberland 2002; Goldberg 2009). These efforts were strongly resisted by the United States, and implementation was hindered by economic austerity, structural adjustment programmes, and resulting cutbacks in primary health services across Latin America (Cohen 2003; Alvaro, Palma, and Dardet 2006). Despite these obstacles, international discourses of reproduction since Cairo have tended to be framed in terms of individual rights. National campaigns to limit fertility are still in place, but with birth rates dropping dramatically the battle to limit population growth was considered largely over. Collective notions of population control and reproductive health have given way to governance through a new – and newly juridical – understanding of individual rights. This discursive formulation has created an opening for competition between the 'right-to-life' of the unborn and the 'reproductive rights' of women.

Changes in reproductive governance must be understood within the context of neoliberalism 'and its discontents' (Bedford 2009; Tsai 2009; Ewig 2010). Neoliberal economic policies have not been uniformly accepted across Latin America; while some countries continue to pursue market-oriented neoliberalism, others retain state-sponsored social supports, and some like Venezuela, Bolivia, Nicaragua and to a lesser extent Ecuador, have joined the counter-hegemonic Bolivarian Revolution (known in Spanish as ALBA; *Alianza Bolivariana para los Pueblos de Nuestra América*). Rivalry between diverse political strategies sometimes extends to reproductive governance, with one party arguing that its policies are more rational (in the sense of humane, efficient, or compassionate) than those of its opponent. Yet neoliberalism has provided a robust breeding ground for shifting political rationalities of reproduction that centre, in part, on moral regimes based in rights claims.

Events in Peru exemplify these regional shifts in reproductive governance. The conservative administration of Alberto Fujimori (1990–2000) was responsible for the clandestine coerced sterilisation of hundreds of thousands of indigenous women in the 1990s, a campaign that prompted a national and international outcry when it became public in 2002 (Coe 2002; Center for Reproductive Rights 2003; Ewig 2010). The election of Alejandro Toledo in 2001 dramatically shifted approaches toward reproductive governance. Two of his actions while in office exemplify his approach. To distance himself from Fujimori, Toledo worked to make state institutions recognise the rights claims of indigenous groups based on notions of collective sovereignty – that is, sovereignty based in the collective identity of indigeneity – instead of individual citizenship. These efforts were a partial response to concerns about the ability of indigenous groups to reproduce themselves in the wake of the sterilisation abuses. Toledo also furthered the interest of the Catholic Church by proposing a legal code that would have required all Peruvian women to register their pregnancies at the time of conception to protect the rights of the unborn. Both of these endeavours declared the sanctity of life from conception and the rights of the unborn. What is most significant in the authors' view is that this shift in governance between the administrations of Fujimori and Toledo – from the limitations of indigenous women's fertility through clandestine population control, to the official attempt at the expansion of state surveillance of all Peruvian's women's reproduction through the rhetoric of rights – took place while Toledo's administration cut back access to reproductive health services including condoms, emergency contraception, and post-abortion hospital care (Coe 2004).

It is important to point out that 'rights' discourses take somewhat different forms in Latin American than in other contexts. The concept of 'human rights', for example, tends to have a more collective valence, especially in countries that witnessed sustained campaigns of disappearance and torture under military dictatorships. Human rights are thus understood to apply to the citizenry broadly conceived, and not only to individuals. The theory of 'natural rights' is also heard more frequently in Latin America, where it refers to divinely given rights that exempt individuals from both state control and the reach of secular law. According to social theorist Margaret R. Somers and legal theorist Christopher N.J. Roberts, natural rights are 'alleged to be morally justified by higher laws of God and nature and possessed universally by individuals' (Somers and Roberts 2008, 289). The concept of natural law is invoked to support the supposedly universal and inalienable rights of embryos and foetuses. Proponents argue that legislators have no right to intervene in the 'natural right' of embryos to be born, because any such intervention would be arbitrary, indefensible, and unnatural. The concept of 'indigenous rights' may trump reproductive rights in places such as Bolivia and Peru, where the new constitution has bestowed rights upon indigenous people, and where women's rights are framed as a movement to eliminate ethnic discrimination and to humanise medical care for all women, rather than on the right to bodily autonomy. This is a very different model than US-based rights claims about the 'choices' to which individual women should be entitled. (Indeed a broader comparison of reproductive governance across the Americas, while beyond the scope of this paper, would be instructive.) The proliferation of diverse rights claims encourages different constituencies to pursue their claims in courts, as actors separate from the state often in conflict with each other (Harvey 2005).

Rights from the beginning: Legislative and juridical shifts

In the post-population environment that emerged after Cairo, several Latin America legislatures have moved to change the point at which juridical rights commence. Argentina, Ecuador, El Salvador, and Peru, for example, have revised their constitutions and civil codes to push juridical rights back from birth to conception. The Catholic and evangelical churches, in concert with the international right-to-life and pro-family movements, have promoted these legislative changes and enshrined the concept of foetal rights, often with the support of leftist political parties. Since 1998 both El Salvador and Nicaragua have enacted total bans on abortion, two of out of only three countries in the world to outlaw the procedure in all instances (Heumann 2007). In addition to legislation, these various institutions have promoted celebrations and symbols of the international pro-life movement (Morgan 1998). March 25 has been adopted as the Day of the Unborn Child, for example, in El Salvador, Ecuador, Argentina, Chile, Guatemala, Costa Rica, Nicaragua, the Dominican Republic, and Peru.

The instantiation of foetal rights in Nicaragua garnered international attention when current president Daniel Ortega helped to create one of the most punitive abortion laws in the world (Goldberg 2009). When Ortega was a leader of the revolutionary Sandinistas, who ruled from 1979–90, the participation and emancipation of women was vital to his political strategy. The Sandinistas never legalised abortion, however, and the war and the resulting perception of under-population reinforced pro-natalist attitudes even prior to 1990 (Howe 2007). After the Sandinistas lost power in 1990, Nicaragua's vocal and well-organised women's groups lobbied heavily for the liberalisation of abortion laws, but the new leaders instead cultivated support from the Catholic Church, which had backed the Sandinistas' earlier struggles under the banner of liberation theology. In 2006, then President Enrique Bolaños signed a law to ban abortion under all circumstances, just a few days before Ortega was to be inaugurated and apparently with his full support. Complications from unsafe abortion were a leading cause of death in Nicaragua throughout the twentieth century, and with this legislation global health officials warned that the new policy was likely to drive maternal mortality ratios even higher (Lakshmanan 2006). The situation in Nicaragua (and in neighbouring El Salvador) has become a flashpoint for global attention to this issue, as several European Union countries threatened to withhold development assistance while global anti-abortion organisations rallied to defend and emulate the measure as upholding the rights of the unborn.[1]

Patterns of reproductive governance surrounding the question of 'when rights begin' show evidence of local stratification, compromises, and constraints in addition to external pressures and inducements. While in some nations organisation and state institutions have successfully mandated the 'right to life', in others, NGOs and political activists have mobilised under a rights-based approach to advance their stated goals of reproductive safety and autonomy. Networking among Latin American and Caribbean feminists has resulted in a number of coordinated initiatives, including the decision to designate 28 September as a 'Day for the Decriminalization of Abortion' (Gómez 2004) and to share ideas for strategies to combat high rates of abortion-induced maternal mortality using the frameworks of 'human rights' and the Millennium Development Goals (PAHO n.d.). Struggles waged since that time have led to changes in Mexico City, which recently legalised abortion during the first trimester, and in Uruguay, where doctors are now allowed to instruct women on safer medical (i.e., pharmaceutically rather than surgically induced) abortion methods that can be performed at home.

Both human rights groups and the Catholic Church actively monitored and participated in producing recent legislative shifts around abortion in Colombia. In 2005, Colombian lawyer Mónica Roa challenged the constitutionality of the abortion prohibition (Article 122 of the Penal Code). She argued not on the basis of a right to privacy (which was the basis of the claim made in *Roe v. Wade* in the United States), but in favour of women's constitutional 'rights to life, health and physical integrity'. Roa had hoped to push for full legalisation but decided a narrower focus on the right to health would be more effective. While human rights activists argued that there were over 450,000 clandestine abortions in Colombia every year and that unsafe abortion was a leading cause of maternal mortality, the Colombian Catholic Church contested these claims, arguing that abortion did not constitute a health problem and thus should remain illegal (Catholic News Agency 2005). In May 2006, Colombia's Constitutional Court declared that 'neither women nor doctors can be penalised for procuring or providing abortions where one of three conditions is met: 1) the pregnancy constitutes a grave danger to the pregnant woman's life or health; 2) the foetus has serious genetic malformations; or 3) the pregnancy is the result of rape or incest'. The court voted 5-3 in favour of partial legalisation despite the opposition of conservative president Alvaro Uribe, who was then in the midst of a re-election campaign (Karsin 2006). His leftist opponent favoured decriminalisation. The Colombian case helped to underwrite an international shift in strategy for progressives, in which proponents utilised the concept of 'health clauses' in their efforts to decriminalise abortion. The moral regime of the right to 'health' was therefore counterposed against that of 'conscience', where both would supposedly support the most supreme value, that is, the right to life itself.

As pointed out earlier in the case of Daniel Ortega in Nicaragua, the political rationalities and newly minted reproductive policies emerging in Latin America do not correspond to prevalent North American assumptions about how leftist or right-wing governments should act (Edelman and Haugerud 2005; Castañeda 2006; Gago 2007; Arditi 2008; Azize Vargas 2009). The election of progressive leaders has not necessarily resulted in progressive reproductive health policies. Some of the most left-leaning administrations in recent years – including those of Hugo Chávez in Venezuela, Rafael Correa in Ecuador, Evo Morales in Bolivia, and Tabaré Vázquez of Uruguay – oppose the liberalisation of abortion policies in their countries, although Lula da Silva in Brazil and Michelle Bachelet in Chile were both on record arguing for the liberalisation of abortion law. Meanwhile, some right-wing administrations, such as those of Alvaro Uribe in Colombia and Felipe Calderón in Mexico, presided over the unprecedented liberalisation of abortion policies. How to explain these seeming contradictions?

Some have cited lingering machismo and entrenched Catholicism (see Friedman-Rudovsky 2007), although anthropologists argue that machismo may be an outmoded concept in light of changing attitudes toward masculinity and gender relations (Dudgeon and Inhorn 2004; Gutman 2007, 9–11). The Catholic hierarchy has indeed been a formidable political force, especially where elite opponents of reproductive rights are educated in schools run by Opus Dei (Vaggione 2010). But reproductive health policies have also been liberalised in countries that are predominantly Catholic. Many Latin American Catholics embrace modern contraception and sterilisation, and fertility clinics are proliferating despite the Church's adamant condemnation of assisted reproduction (Roberts 2006). And although abortion is illegal in most Latin American countries, it is primarily Catholic women who account for some of the highest abortion rates in the world (Browner 1976; Htun 2003; Scrimshaw 1985). Many leftist political parties, it should be noted, owe their existence to the Catholic Church, which supported them during military dictatorships. The *quid pro quo* for their support may be that leftist leaders do not challenge the Church's condemnation of abortion. It is clear that Catholic practice in Latin America is far from monolithic (Voekel 2002; Vaggione 2005; Roberts 2006). An understanding of these ostensibly divergent trends cannot rely on culturalist explanations such as gender or religion alone, without taking global and regional economic transformations into account.

Besides producing rights bearers that stand in opposition to each other, rights-based reproductive governance supports neoliberal agendas in other ways as well, by producing citizens who have the 'right to choose' (that is, to consume) privatised medical services. Wealthy women, for example, find it relatively easy to locate doctors willing to perform abortions in the private sector, even though it is illegal. Because safe clinical abortion is readily available for women with economic resources, it has been difficult to convince middle- and upper-class women and their partners to mobilise around safe abortion as a cause (Htun 2003). While states such as Brazil and Ecuador have enacted free maternity laws under the banner of the right to reproductive health, in practice there has been significant retrenchment in access to primary health care across the hemisphere, including maternity and prenatal services. This retrenchment is often couched in terms of enhancing the consumer's 'right to choose' (Burrows 2008). The retraction of state commitments to public health has meant the rapid expansion of privatised medical care, leading to high C-section rates and a burgeoning infertility industry (Roberts 2012). Assisted reproductive technologies, including in-vitro fertilisation, flourish in Latin America despite the Catholic Church's total condemnation and the fact that the legal rights-based focus on conception situates IVF in ambiguous legal territory (Center for Reproductive Rights 2004). Ecuador provides a painful example of how divergent rights-based reproductive policies congeal in the production of new consumers. Some of the poorer women who end up in IVF clinics in Ecuador became infertile after an unsafe abortion. One could argue in this case that state institutions that govern reproduction in the name of the 'rights of the unborn' have directly produced consumers 'with rights' for privatised reproductive medicine (Roberts 2012).

A population of migrants

Reproductive governance provides a way to trace how different institutional actors are shaping the management of national populations in the post-Cold War era. Indeed the very notion of 'population' itself might be undergoing shifts, given that life and reproduction are conceptualised mostly in terms of individual rights. Now population concerns frequently congregate around migrations to North America and Europe and between Latin American nations. Here, older concerns about overpopulation resurface in the contradictory figure of the migrant. The departure of the émigré leaves a gap, resulting in lower economic productivity and skewed sex ratios in the country of origin, as well as a significant influx of remittances. These same immigrants are often portrayed as a threat to the body politic in their destination countries, where they are portrayed as devouring scarce resources, bearing disease, breeding indiscriminately, and otherwise challenging the boundaries of decency and citizenship. Tensions between these contradictory portraits can be seen in concerns over pregnant Nicaraguan women who migrate to Costa Rica, where they are depicted as exploiting the resources and goodwill of their main benefactor, the Costa Rican state (Goldade 2007). While these migrants are increasingly demonised as sucking resources form their host nation, the foetus is increasingly made into a citizen worthy of protection and juridical rights (although not state provided social services). Foetal citizens with juridical rights are brought into social being by policies that prohibit doctors from prescribing medications or treatments that might restrict or interfere with conception, implantation, or gestation. In April 2008, the Chilean Constitutional Court overturned the government's new policy making emergency contraception available, free and on demand, to all women over age 14 who requested it. This ruling effectively privileges foetal citizens over women, by prohibiting the state from prescribing medications that would prevent unwanted conception. This contestation produced two of the duelling figures in the new rationality of reproductive governance; the immigrant-as-resource-depleter to whom rights can be denied, versus the rights-bearing foetus-as-citizen who takes nothing from the neoliberal state.

Reproductive governance

To paraphrase anthropologists Faye Ginsburg and Rayna Rapp, a focus on the governance of reproduction belongs at the centre of social theorising about contemporary Latin American politics and economies (Ginsburg and Rapp 1995, 1). Characterising reproductive governance in Latin America's varied states requires that anthropologists pay close ethnographic attention to what politicians, NGOs, churches, and women's groups gain, rhetorically and politically, by claiming to be concerned about 'the rights of the unborn', 'the rights of women', 'reproductive rights', 'natural rights', 'indigenous rights', and 'consumer rights'. It requires attention to the new spate of laws designed to increase surveillance, regulation, and prosecution, and to how reproductive governance is meant to enact an ideal political imaginary. How, in other words, might reproductive governance produce certain kinds of subjects and effects?

One of the authors' main methodological concerns is that a fine-grained ethnographic focus on nations be kept in conversation with an analysis of regional, hemispheric and transnational political and economic processes. The country-specific examples offered in this essay provide important instances of local possibility and constraint, but the scholarship on reproduction and population has often been limited by 'research questions based on national histories' rather than pan-national or global questions (Bashford 2006, 173). Comparative and collaborative ethnography in conjunction with cross-regional historical and political economic analysis can make more apparent how political and economic processes enact certain subject positions that can seem so ahistorical, such as irresponsible mothers or the unborn, or the concept of rights itself. Research across borders shifts the focus towards understanding the deployment of rights rhetoric and changing rationalities of reproduction throughout the region. The concept of reproductive governance should be applicable to other settings where shifting political rationalities call forth new moral regimes and rights-based actors in the domain of reproduction. This kind of research allows for richer comprehension of how new laws, social movements, moral incitements and economic inducements are working to police, regulate, and coerce reproductive bodies and to produce self-controlled subjects who will embody contemporary forms of governance.

Reproductive governance in Latin America has always reproduced social distinctions, identities, alliances and produced subjects and citizens, at times solidifying ethnic boundaries, while at others producing powerful political alliances, complacent mothers, and industrious wage-earners. At this historical juncture an analytics of reproductive governance can elucidate how the national and transnational configuration that produces the rhetoric of rights makes new kinds of reproductive actors and fits within larger social movements and the geopolitical and economic calculus of Latin American nation-states.

Acknowledgements

The authors would like to thank Silvia De Zordo and Milena Marchesi for the invitation to participate in this special issue as well as their guidance in revising the piece. Roberts would like to thank Diane Nelson, Michelle Murphy and S. Lochlann Jain of Oxidate, as well as Rebecca Hardin, Mark Padilla and Matt Hull for their generous comments and suggestions on this text and their help conceptualising reproductive governance. The NSF and the Wenner Gren Foundation funded Elizabeth Robert's ethnographic research, which informed her analysis. Lynn Morgan is grateful to the School for Advanced Research and Mount Holyoke College for supporting this research, and to Charles Briggs, Clara Mantini-Briggs, Chris Teuton, Sherry Farrell Racette, and James Trostle for their comments and camaraderie. Institutional ethics review was not required for this research.

Conflict of interest: none.

Note

1. In an internationally publicised case that occurred a week after the new legislation was signed into effect in Nicaragua, an 18-year old woman died of septic shock after an illegal abortion because doctors were afraid to give her antibiotics lest they be accused as accomplices.

References

Alvarez, Sonia. 1998. Latin American feminisms 'go global': Trends of the 1990s and challenges for the new millennium. In *Culture of Politics, Politics of Cultures: Re-Visioning Latin American Social Movements*, eds. E.D. Sonia, E. Alvarez and Arturo Escobar, 293–324. Boulder, CO: Westview.

Alvaro, Franco Giraldo, Marco Palma, and Carlos Alvarez Dardet. 2006. Efecto del ajuste estructural sobre la situación de salud en América Latina y el Caribe, 1980–2000. *Revista Panamericana de Salud Pública* 19, no. 5: 291–9.

Aramburú, Carlos. 1994. Is population policy necessary? Latin America and the Andean countries. *Population and Development Review* 20 (Supplement): 159–78.

Arditi, Benjamin. 2008. Arguments about the left turns in Latin America: A post-liberal politics? *Latin American Research Review* 43, no. 3: 59–81.

Azize Vargas, Yamila. 2009. La izquierda y el aborto en América Latina y el Caribe. *Revista Mujer Salud RSMLAC* 1: 54–8.

Bashford, Alison. 2006. *Medicine at the border: Disease, globalization and security, 1850 to the present*. Basingstoke, UK, New York: Palgrave Macmillan.

Bashford, Alison, and Philippa Levine. 2010. *The Oxford handbook of the history of eugenics*. Oxford: Oxford University Press.

Bedford, Kate. 2009. *Developing partnerships: Gender, sexuality, and the reformed World Bank*. Minneapolis: University of Minnesota Press.

Browner, Carole. 1976. Poor women's fertility decisions: Illegal abortion in Cali, Colombia. PhD Thesis in anthropology, University of California Berkeley.

Burrows, Jamie. 2008. Inequalities and healthcare reform in Chile: Equity of what? *Journal of Medical Ethics* 34, no. 13.

Cardarello, A. 2012. The right to have a family: 'legal trafficking of children', adoption and birth control in Brazil. *Anthropology & Medicine* 19, no. 2: 225–40.

Carranza, Maria. 2007. The therapeutic exception: Abortion, sterilization and medical necessity in Costa Rica. *Developing World Bioethics* 7, no. 2: 55–63.

Castañeda, Jorge. 2006. Latin America's left turn. *Foreign Affairs* 85, no. 3:28–43.

Catholic News Agency 2005. Family institute in Colombia refutes statistics on abortion. Vol. 2008.

Center for Reproductive Rights 2003. *Silence and complicity: Violence against women in Peruvian public health facilities*. New York: Center for Reproductive Rights.

Center for Reproductive Rights 2004. Center joins couples' legal battle against Costa Rica's IVF ban. New York: Center for Reproductive Rights.

Coe, Anna-Britt. 2002. Informing choices: Expanding access to emergency contraception in Peru. Washington, DC: Center for Health and Gender Equity.

Coe, Anna-Britt. 2004. From anti-natalist to ultra-conservative: Restricting reproductive choice in Peru. *Reproductive Health Matters* 12, no. 24: 56–69.

Cohen, Susan A. 2003. Bush administration isolates U.S. at international meeting to promote Cairo agenda. *Guttmacher Report on Public Policy*, 6, no. 1: 3–5.

Corrales, Javier. 2010. Latin American gays: The post-left leftists. *Americas Quarterly Online*. http://www.americasquarterly.org/gay-rights-Latin-America.

Donzelot, Jacques. 1979. *The policing of families*. New York: Pantheon Books.

Dudgeon, Matthew, and Marcia Inhorn. 2004. Men's influences on women's reproductive health: Medical anthropological perspectives. *Social Science & Medicine* 59, no. 7: 1379–95.

Edelman, Marc, and Angelique Haugerud. 2005. *The anthropology of development and globalization: From classical political economy to contemporary neoliberalism*. Malden, MA: Blackwell.

Ewig, Christina. 2010. Second-wave neoliberalism: Gender, race, and health sector reform in Peru. University Park, PA: Pennsylvania State University Press.

Fassin, Didier. 2007. Humanitarianism as a politics of life. *Public Culture* 19, no. 3: 499–520.

Foucault, Michel. 1990. *The history of sexuality*. New York: Vintage Books.

Foucault, Michel. 2003. *Society must be defended: Lectures at the College de France, 1975–76*. New York: Picador.

French, William, E., and Katherine Elaine Bliss. 2007. *Gender, sexuality, and power in Latin America since independence*. Lanham, MD: Rowman & Littlefield.

Friedman-Rudovsky, Jean. 2007. Abortion under siege in Latin America. *Time Magazine*, August 9: http://www.time.com/time/world/article/0,8599,1651307,00.html

Gago, Verónica. 2007. Dangerous liaisons: Latin American feminists and the left. *NACLA Report on the Americas* 40, no. 2: 17–9.

Gibson-Graham, J.K. 1996. *The end of capitalism (as we knew it): A feminist critique of political economy*. Cambridge, MA, Oxford: Blackwell Publishers.

Ginsburg, Faye, D., and Rayna Rapp. 1995. *Conceiving the New World Order: The global politics of reproduction*. Berkeley, CA: University of California Press.

Goldade, Kate. 2007. Reproducción transnacional: la salud reproductiva, las limitaciones y las contradicciones para las migrantes laborales nicaragüenses en Costa Rica. In *El Mito Roto: Inmigración y Emigración en Costa Rica*, ed. C.S. García, 233–60. San José: UCR Press.

Goldberg, Michelle. 2009. *The means of reproduction: Sex, power, and the future of the world*. New York: Penguin Press.

Gómez, Adriana. 2004. 28 de septiembre: Por la despenalización del aborto: La mujer decide, la sociedad respeta, el Estado garantiza. Red de salud de las mujeres latinoamericanas y del caribe.

Goodale, Mark, and Sally Engle Merry. 2007. The practice of human rights: Tracking law between the global and the local. Cambridge: Cambridge University Press.

Gutman, Matthew. 2007. Fixing men: Sex, birth control, and AIDS in Mexico. Berkeley, CA: University of California Press.

Haberland, Nicola, and Diana Measham. 2002. *Responding to Cairo: Case studies of changing practice in reproductive health and family planning*. New York: Population Council.

Hartmann, Betsy. 1999. *Reproductive rights and wrongs: The global politics of population control*. Boston: South End Press.

Harvey, David. 2005. *A brief history of neoliberalism*. Oxford: Oxford University Press.

Heumann, Silke G. 2007. Abortion and politics in Nicaragua: The women's movement in the debate on the Abortion Reform Law, 1999–2002. *Culture, Health & Sexuality* 9, no. 3: 217–31.

Howe, Cymene. 2007. Gender, sexuality, and revolution: Making histories and cultural politics in Nicaragua, 1979–2001. In *Gender, sexuality, and power in Latin America since independence*, ed. W.E. French and Katherine Elaine Bliss, 230–60. Lanham, MD: Rowman & Littlefield.

Htun, Mala. 2003. Sex and the state: Abortion, divorce, and the family under Latin American dictatorships and democracies. Cambridge, New York: Cambridge University Press.

Karsin, Nicole. 2006. Abortion adds to Colombia's election turmoil. *Women's E-News*. http://womensenews.org/story/campaign-trail/060525/abortion-adds-colombias-election-turmoil.

Krause, Betsy, and Milena Marchesi. 2007. Fertility politics as 'social viagra': Reproducing boundaries, social cohesion, and modernity in Italy. *American Anthropologist* 109, no. 2: 350–62.

Lakshmanan, Indira A.R. 2006. Nicaragua abortion ban called a threat to lives. *Boston Globe*, November 26.

Lane, Sandra. 1994. From population control to reproductive health: An emerging policy agenda. *Social Science & Medicine* 39, no. 9: 1303–14.

Leite, Iúri da Costa, Neeru Gupta, and Roberto do Nascimento Rodrigues. 2004. Female sterilization in Latin America: Cross-national perspectives. *Journal of Biosocial Science* 36: 683–98.

McWhorter, Ladelle. 2004. Sex, race, and biopower: A Foucauldian genealogy. *Hypatia* 19, no. 3: 38–62.

Mooney, Jadwiga, and E. Pieper. 2009. *The politics of motherhood: Maternity and women's rights in twentieth-century Chile*. Pittsburgh, PA: University of Pittsburgh Press.

Morgan, Lynn. 1998. Ambiguities lost: Fashioning the fetus into child in Ecuador and the United States. In *Small wars: The cultural politics of childhood*, ed. N. Scheper-Hughes and C.F. Sargent, 58–74. Berkeley, CA: University of California Press.

Morgan, Lynn, and Elizabeth F.S. Roberts. 2009. Rights and reproduction in Latin America. Anthropology News, 50: 12–16.

Necochea López, Raúl. 2008. Priests and pills: Catholic family planning in Peru, 1967–1976. *Latin American Research Review* 43, no. 2: 34–56.

PAHO, Pan American Health Organization. n.d. Millennium development goals. regional situation analysis. *Population Reports*.

Roberts, Elizabeth F.S. 2012. *God's laboratory: Assisted reproduction in the Andes*. Berkeley, CA: University of California Press.

Roberts, Elizabeth F.S. 2006. God's laboratory: Religious rationalities and modernity in Ecuadorian *in vitro* fertilization. *Culture Medicine and Psychiatry* 30, no. 4: 507–36.

Scrimshaw, Susan C.M. 1985. Bringing the period down: Induced abortion in Ecuador. In *Micro and macro levels of analysis in anthropology: Issues in theory and research*, ed. B.R. DeWalt and P.J. Pelto, 121–46. Boulder, CO: Westview Press.

Somers, Margaret R., and Christopher N.J. Roberts. 2008. Toward a new sociology of rights: A genealogy of 'buried bodies' of citizenship and human rights. *Annual Review of Law and Social Science* 4: 385–425.

Stoler, Ann Laura. 1995. *Race and the education of desire: Foucault's history of sexuality and the colonial order of things*. Durham, NC: Duke University Press.

Tsai, Thomas and John Ji. 2009. Neoliberalism and its discontents: Impact of health reforms in Chile. *Harvard International Review* 31, no. 2: 32–36.

Vaggione, Juan Marco. 2005. Reactive politicization and religious dissidence: The political mutations of the religious. *Social Theory and Practice* 31, no. 2: 1–23.

Vaggione, Juan Marco. 2010. Evangelium Vitae today: How conservative forces are using the 1995 Papal Encyclical to reshape public policy in Latin America. *Conscience* 31, no. 2: 23–30.

Voekel, Pamela. 2002. *Alone before God: The religious origins of modernity in Mexico*. Durham, NC: Duke University Press.

Wilson, Ara. 2004. *The intimate economies of Bangkok: Tomboys, tycoons, and Avon ladies in the global city*. Berkeley, CA: University of California Press.

Index

abortion: in Brazil 8, 72, 74, 82; in Italy 35–8, 41–6; in Latin America 12, 106, 111–13; in Pakistan 58, 65; in Poland 18, 21–2, 28
adoption of children 7, 106
aging population 18
Alianza Bolivariana para los Pueblos de Nuestra América (ALBA) 110
Al-Musnad, Muhammad bin Abdul-Aziz 59
assisted reproductive technologies (ART) 37, 41–2, 113; *see also* fertility treatment
authoritarian attitudes 82
Ayub Khan 53

'baby bonuses' 38
Bachelet, Michelle 113
Beatriz, Dr 75
Beijing conference on women (1995) 109
Benedict XVI, Pope 40, 42
Berlusconi, Silvio 38
Bhutto, Zulfiqar Ali 53
bio-medicine 13
biopolitics and biopower 2, 4–5, 19, 30–1, 36, 40–1, 54–5, 66, 71–3, 82–4, 107–8
birth spacing 62–3, 67
Blangiardo, Gian Carlo 44
Bolaños, Enrique 111
Bolivia 110–11
Brazil 7–13, 71–84, 106; Constitution 72–3; Law on Family Planning (1996) 74; Parliamentary Commission (1991) 73

caesarean section 8, 73
Cairo conference on population and development (1994) 109
Calderón, Felipe 113
Camusso, Susanna 43
Cardarello, Andréa 4, 7; *contributing author*
Catholic Church 6–9, 12, 17–21, 28–31, 36, 40–2, 47, 80, 106–13
Celia, Dr 76–81
Chávez, Hugo 112–13
childcare 23–4
childlessness 26, 64
Chile 114
Ciampi, Azeglio 43
citizenship 46
coitus interruptus 44, 58
Cold War 109
Colombia 12, 112
contraception: in Brazil 71–83; in Italy 37–8, 42–4; in Latin America 109, 113–14; in Pakistan 53–67; in Poland 21–2, 28; male partners' involvement in 80–3; risks involved in use of 65; side-effects of and complaints about 76–82; social acceptability of 66
Correa, Rafael 112–13
Costa Rica 106, 114
Coutinho, E. 71–2
culture, concept of 82

demographic science 5, 9–10, 37–8
De Zordo, Silvia 4, 7–8, 11; *co-editor and contributing author*
disciplining measures 6, 8, 19
discrimination in employment 24–5, 29
Dominican Republic 12

Ecuador 113
education 8
Ehrlich, Paul 4
El Salvador 12, 111
embryo protection 37, 41–2, 110
empiricism 5
ethnographies of biopolitics 4–5
ethnopolitics 66
eugenic programs 10, 72, 74

Fábio, Dr 74–6
family planning 6–8; in Brazil 10–11, 71–6, 82–3; in Italy 42, 44; in Pakistan 53–67; in Poland 17–21; Islamized 9, 54–67; physicians' view of 74–6, 83; as a *rational practice*, as a *civic duty* or as *right* 71–3; as a social responsibility 82; 'tactical' use of 82

Family Planning Association of Pakistan (FPAP) 55–67; Empowerment of Adoles-cents project 60; Girl Child Program 60
family size *see* multi-child families; second children; 'smaller families' model
family traditions 63
fascism 37–8, 75
Fassin, Didier 2, 14, 73, 83, 106
fatwas 57–60, 65–6
Fazio, Ferruccio 44
feminism and feminist groups 11–13, 22, 28, 38–9, 43, 74, 112
Ferreri, Silvia 43
fertility politics 4
fertility rates 1–6, 9–10; in Brazil 72–4; in Italy 35–9, 43–6; in Latin America 109, 113; in Pakistan 54–7; in Poland 17–20
fertility treatment 41, 46, 80–1; *see also* assisted reproductive technologies
foetal rights 42, 106, 110–11, 114
Foucault, Michel 2–4, 10, 13–14, 19, 30, 36, 106–8
Fujimori, Alberto 110

Gal, Susan 3, 43
Gazeta Wyborcza 18
gender inequities 18–19, 30–1, 83
genealogical research 5
Gilgit-Baltistan region 9, 53–67
Gine-Salvador clinic 73–6, 81
Ginsburg, Faye 3, 114
Giovinardi, Carlo 39
Good, Byron 83
Greenhalgh, S. 38

Hacking, Ian 2
Hadith 57–62
Hanafin, Patrick 36
Harvey, David 107
health care 113
Henzler, M. 18
Horn, D.G. 37
human rights 7, 107, 110, 112
Humanae Vitae (papal encyclical, 1968) 40

immigrants *see* migrant populations
indigenous peoples 110–11
inequality 14, 22; *see also* gender inequities
Inter-American Commission on Human Rights 107
in-vitro fertilization (IVF) 41, 106, 113
Islam and Islamic doctrine 8–9, 53–5, 60–3, 66–7; *see also* Muslim communities
Islamic reform movements 63, 66–7
Islamist movements 54
Islamization *see* family planning, Islamized

Italian Gynaecological and Obstetrical Society (Sigo) 40
Italy 6–13, 35–47; pro-natalist policies 37–9

Jeffrey, P., R. and C. 55
John Paul II, Pope 28
Johnson-Hanks, Jennifer 5–6

Kanaaneh, Rhoda 46
Kligman, Gail 3, 43
Krause, Elizabeth 10, 39; *contributing author*
Kurkiewicz, Katarzyna 25

Latin America 12, 105–15
Latour, B. 80
Lopez, I. 84
Lula da Siva, Luiz Inácio 113

McDonald, P. 18
machismo 80, 83, 113
Malthus, Thomas 2–3; *see also* neo-Malthusian ideas
Marchesi, Milena 4, 9–13; *co-editor and contributing author*
Martin, Emily 77
Marx, Karl 2–3
Mexico City 12, 112
middle-class identity 29–30
migrant populations 6–7, 10, 35–9, 43–6, 56, 106–9, 114
Milan conference on fertility (2007) 35–6
Milan May Day Parade (2007) 46
Mishtal, Joanna 4, 6, 10, 12; *contributing author*
moral regimes 13, 106–8, 115
Morales, Eva 112–13
Morgan, Lynn M. 10, 12–13
mortality: infant 72–3; maternal 72–4, 111–12
motherhood, attitudes to 23, 26–30
mothers-in-law, influence of 63, 65
multi-child families 26–7
Muslim communities 29
Mussolini, Benito 2

natural rights and natural law 110
neoliberalism 1, 6, 11, 14, 21–2, 30–1, 43, 105–10, 113
neo-Malthusian ideas 7–9, 71, 75, 82
Nicaragua 12, 106, 111–12, 114

Ong, A. 36
Ortega, Daniel 111–12
'otherness' 83
overpopulation 4, 38, 109, 114

Pakistan 8–12, 53–67; Ministry of Population Welfare 55–8, 61–4

parental rights 7
patriarchy 30
patronising attitudes 82
Paul VI, Pope 40
Pawlicki, J. 18
Peru 110–11
Phulwari, Mualana Alhaaj Shah Muhammad Jaffar 58–9
Poland 6, 9–13, 17–31, 36; Profamily Program 19, 26, 30–1
Polityka 17–18
poor laws 2–3
population control policies and initiatives 53–4, 65–6, 106, 109–10
population projections 4, 44
postponement of childbearing 17–18, 22–5, 28–9, 71
poverty associated with criminality 106
power, theory of 3; *see also* biopolitics and biopower
'Precious Pearl' brief 59–60
pregnancy: benefits attached to 65; and employment 25; risks involved in 63–4
preventive medicine 72
Princeton European Fertility Project 5
privatization 13–14, 21; of medical care 113
Prodi, Romano 38

quality of life 57–8, 72
Qur'an, the 61

Rabinow, Paul 13, 54
Rapp, Rayna 3, 114
'rational reproduction' 59–61, 66–7, 72
rationality in human behavior 13–14, 28–31, 38–9
Reher, D. 18
'replacement anxiety' 35, 39–40
'reproductive governance' 105–15
'reproductive Westernization' 38–9
'responsible reproduction' 39–40, 46
right to life 112
'rights' discourse 74, 107–15; *see also* human rights; women's rights
Roa, Mónica 112
Roberts, Christopher N.J. 110
Roberts, Elizabeth F.S. 10, 12–13
Roma people 6, 26
Rose, Nikolas 13, 54, 66, 73, 82
Rzeczpospolita 29

Salvador da Bahia 71–4, 82–4
Sanabria, Emilia 82
Sandanistas 111
Schneider, Jane and Peter 26–7
second children 23, 26, 82
sectarian affiliation 62
sectarian violence 62, 65
secularism 61
sex education 19, 21, 27
sexuality 107–8
'smaller families' model 57–9, 63–7
Smith, Adam 2
Solidarity 21
Somers, Margaret R. 110
sovereignty, collective 110
state socialism 6, 20–1
sterilization: in Brazil 7–8, 11, 73–4, 81–2; in Latin America 110, 113; in Pakistan 58
stigmatization of women 6–9, 26–7, 30, 65, 71, 82
surplus population 2–3
surveillance of reproductive behaviors 6, 19, 110

Tablighi Jamaat 61–2
'tempo effect' on fertility 18, 30
Toledo, Alejandro 110
Treves, A. 38
Trondman, Mats 5
tubal ligation 8–11, 72–4, 80–3, 106
Turco, Livia 38
'two child' model 57

ulema, influence of 53–4, 57–8, 67
Underwood, C. 54
Uribe, Alvaro 112–13
Uruguay 112

Varley, Emma 4, 8–9; *contributing author*
vasectomies 73–4, 80–1
Vásquez, Tabaré 112–13
'vitapolitics' 35–7, 40–1, 46
Vittori, Giorgio 40
Vlk, Miroslaw 29

Wałęsa, Lech 17, 21
Wall Street Journal 36
Willis, Paul 5
women, characterisations of 62
women's rights 22, 30, 110–12
World Bank 22

Zia Ul-Haq 53